Theory of the optically induced forces

V. P. TORCHIGIN

ISBN: 1511954477
ISBN-13:9781511954471

CONTENTS

Introduction

An optically induced force (OIF) is the force that acts on an optical medium from the side of the light where it is propagating. In accordance with the third Newton law, the momentum of the light should changes. However, a magnitude of the momentum of light in matter is unknown till now. A century has now passed since the origins of the Abraham-Minkowski controversy pertaining to the correct form of optical momentum in matter. There have been extensive debates about the correct expression of the momentum density of electromagnetic waves in linear media for more than 100 years since the original papers of Minkowski [Minkowski 1909] and Abraham [Abraham 1910], and even so there is still some confusion or at least disagreement among authors. In accordance with Minkowski, momentum of light in an optical medium increases by n times as compared with the momentum of the same light in free space where n is the refractive index of the medium. In accordance with Abraham, momentum of light in an optical medium decreases by n times.

Over the years, many theoretical arguments have been advanced and experimental evidence provided in support of one theory over the other, but the controversy persists. At present, this controversy is considered as one of unsolved problem of the theoretical physics.

Experiment and theory have been used in attempt to resolve the debate. There is the recent simplest experiment by She *et al* [She 2008] where the slight deformations observed upon transmission of a light through a short length of a silica glass fiber. Authors explain these deformations by action of optically induced force (OIF) that arises because the radiation of light from the nano-fiber is accompanied by a change of the momentum of light. Relating the shape and magnitude of these deformations to the momentum of the light inside and outside the fiber, authors concluded that the momentum of light in fiber corresponds to the Abraham approach.

Immediately several comments appeared where consideration of authors about a change of the momentum of light responsible for observed deformation are questioned. Brevik in his comments objects that the force responsible for the deformations is produced due to an inhomogeneity of the optical medium located in an electrical field of light wave [Brevik 2009]. Mansuripur offers to take into account the Lorentz force and to take into account not only the electromagnetic momentum but also the mechanical one [Mansuripur 2009]. She *et al* in their reply show that some Mansuripur objections are groundless [She 2009]. Papers with more detail analysis of OIF in the She *et al* experiment has been published later. However, an agreement between debaters about phenomenon that are responsible for the observed motion of the fiber has not been achieved. In the same time, a conclusion derived by She *et al* from their experiment has already mentioned by theorists as an additional argument that the momentum of light in an optical medium corresponds to the Abraham form.

At present, it is not clear what kind of forces ought to be taken into account and who of the debaters is right. Having studied these publications, we can conclude that there is no generally accepted notion either about physical phenomena responsible for a rise of OIF or about their interrelation and interaction in the simplest device used in the She *et al* experiment. Moreover, there is no agreement about kinds of OIF that ought to be taken into account for explanation of the effect. As a result, there is no generally accepted opinion about a conclusion that can be derived from the She *et al* experiment. It is not clear in advance, either all known kinds of OIF ought to be taken into account simultaneously or they are interdependent and some kinds of OIF describe the same force but different terms and notions are used. Besides, there is no confidence also that all kinds of OIF are known.

The state of the art of the OIF theory can be illustrated by the following example. It is well known that a light wave propagating normally to the plane mirror from free space produces the positive pressure on the mirror. But is unknown till now what is the pressure produces the same light wave on a plane boundary of an optical medium. In accordance with Poynting (1905) the pressure is negative. In accordance with Mansuripur (2010) the pressure is equal to zero. In accordance with Loudon (2002) the pressure is positive. We have shown that the Poynting was right.

We presents a solution of this problem that has been published in leading international scientific physical journals in 2013-2014. Of these papers are presented on Appendix.

Grounds for calculation of optically induced forces

Above all things, it is worthwhile to note that our notion about the momentum of light is taken from mechanics where this notion was introduced several centuries ago for a body of mass M moving at speed v as a product of Mv. In accordance with Newton, the momentum characterizes the "quantity of motion". A reason of a change of the momentum p is a force f and in accordance with the second Newton law $dp/dt=f$. In a closed system (one that does not exchange any matter with the outside and is not acted on by outside forces), the total momentum is constant. This fact, known as the law of conservation of momentum, is implied by Newton's laws of motion. [Landau Mechanics 1976]. A magnitude of the momentum of a light pulse propagating in vacuum is generally accepted and is given by $\mathscr{E}_{pulse}/c$, where $\mathscr{E}_{pulse}$ is the energy of the light pulse.

If a continuous plane light wave is considered, its momentum is equal to infinity. In this case, the momentum of the wave propagating through a cross-section of unit area per unit time is considered. This momentum is equal to the momentum density flux (MDF). MDF of a continuous plane light wave propagating in vacuum at speed c is given by

$$M = (W_0 c)/c = W_0 \ [\text{J/m}^3] \tag{1}$$

where

$$W_0 = \varepsilon_0 E_0^2/2 \tag{2}$$

is the energy density of the light wave, E_0 is the amplitude of the strength of the alternate electrical field of the light wave. We will call this MDF by the electromagnetic MDF. A mechanical pressure P applied to a body transmits to the body the mechanical MDF equal $P \ [\text{N/m}^2 = \text{J/m}^3]$.

There are optically induced forces (OIF) produced by the light propagating in an optical medium. As a result, the light interacts with matter (an exchange of the momentums between the light and matter takes place). The law of the conservation of the momentums and the third Newton law are valid at this interaction. As a result, each OIF changes the mechanical MDF of matter. In turn, a counterpart of the OIF (COIF) that arises in accordance with the third Newton law changes the electromagnetic MDF of light. Thus, each interaction is accompanied a redistribution between mechanical and electromagnetic MDFs. A sum of these MDFs is not changed. Thus, OIF is responsible for a change of the mechanical momentum and COIF is responsible for a change of the electromagnetic momentum. Usually, relations between electromagnetic and mechanical MDFs before interaction are known. The mechanical momentum of any light wave in free space is equal zero. Having known a distribution of OIF in space and time, a behavior of the mechanical and electromagnetic MDFs in space and time can be calculated.

The presented below theory is based on the following three facts besides Newton's laws.

1. The MDF of a light wave in free space is equal W_0 where W_0 is the energy density of the light wave. This fact is known since time of Maxwell and was confirmed experimentally by Lebedev at the beginning of the 20th century. Since in accordance with the second Newton law the pressure produced by an electromagnetic wave on an object is equal to a difference between the input and output MDFs, we can conclude that the pressure produced by an electromagnetic wave on an ideal reflector at normal reflection

from it is equal to $2W_0$. In this case the input MDF is equal to W_0 and the output MDF is equal $-W_0$.

2. The momentum of a light pulse propagating in a homogeneous linear lossless dispersionless optical medium decreases by n times as compared with the momentum of the same light pulse propagating in free space. This fact follows from the known Balazs thought experiment which description is presented recently in many publications [Balazs 1953], [Jones 1978], [Mansuripur 2010]. There is no assumption about an origin and kinds of OIF responsible for the decrease of the momentum in the optical medium.

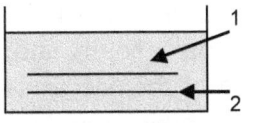

Fig. 1. Plane optical resonator filled with a liquid dielectric

3. The MDF of a plane continuous electromagnetic wave propagating in the same homogeneous linear lossless dispersionless optical medium increases by n times as compared with the momentum of the same wave propagating in free space. This fact is confirmed both real experiments [Jones 1954], [Jones 1978] and thought experiments [Torchigin 2012], [Barnett 2010] where no assumption about an origin and kinds of OIF responsible for the increase of the MDF in the optical medium is done.

Evidence in favour of the Minkowski form of the momentum of light in matter

Momentum of light in a plane resonator

Let us consider a plane optical resonator filled by a liquid dielectric consisting of two plane ideal reflectors as is shown in Fig.1. Let the distance between the reflectors be equal to $d=\lambda/n$, where $\lambda = 2\pi c/\omega$ is the wavelength of light in free space, n is the refractive index. Let there be a resonance at frequency ω and the light wave is reflecting in serial from the reflectors. The force applied to the reflector is equal $F=2p$ where p is the momentum of the light reflected from the reflector per unit time.

Let us decrease slowly the distance by $\Delta d=\Delta\lambda/n$. In this case the resonance is preserved and the wavelength in the resonator decreases from λ/n to $\lambda/n-\Delta d/n = (\lambda/n)(1-\Delta\lambda/\lambda)$. As a result, the frequency of light increases from ω to $\omega/(1-\Delta\lambda/\lambda)^{-1}$ and is independent on n. The frequency of each photon within the resonator increases by the same factor $(1-\Delta\lambda/\lambda)^{-1}$. Since the energy of a photon is proportional to its frequency, the energy E of the light within the resonator increases by the same factor and, therefore, $\Delta E = E(1-\Delta\lambda/\lambda)^{-1}$. As is seen, the increase of the energy is independent on the refractive index n. On the other hand in accordance with the law of conservation of the energy $\Delta E = F \Delta d = 2p \Delta\lambda/n$, where p is the momentum of light that reflects from the reflector of the resonator per unit time. Since ΔE is independent on n, the momentum of light p is proportional to n. Thus, the momentum of light in the resonator corresponds to the Minkowski form.

Momentum of the whispering gallery wave

Let us consider a metallic hollow cylinder filled by the liquid dielectric of the refractive index n. Let the whispering gallery wave (WGW) is propagating along the inner side in such a way that propagation along the axis of the cylinder is absent. There is the resonator of the whispering gallery wave in this case. Let there be the resonance for a certain waveguide mode. Let us decrease slowly and gradually the radius of the cylinder by $\Delta r/n$. In this case the waveguide mode is preserved, but its propagation constant is increased from $\omega n/c$ to $\omega n/c (1-\Delta r/R)^{-1}$ where R is the radius of the cylinder. The resonance frequency ω is increased by the same factor $(1-\Delta r/R)^{-1}$. Since the energy stored in the resonator is proportional to the resonance frequency, the energy is increased by the same factor. As is seen, the energy E is independent on n.

On the other hand, an increase of the energy within the cylinder volume $\pi R^2 h$ is given by $\Delta E = 2\pi RhP(\Delta r/n)$ where h is the height of the volume, P is the pressure produced by the light wave on the side surface of the volume. As we have shown, ΔE is independent on n and, therefore pressure P is proportional to n. The pressure is produced due to a change of the momentum of WGW wave circulating within the volume and, therefore, the momentum of light is proportional to the refractive index n.

Snell law

Let us first consider the pressure produced by a light wave propagating in a planar dielectric waveguide of finite width shown in Fig. 2a where the total internal reflection takes place. In this case the z-component of pressure produced by the

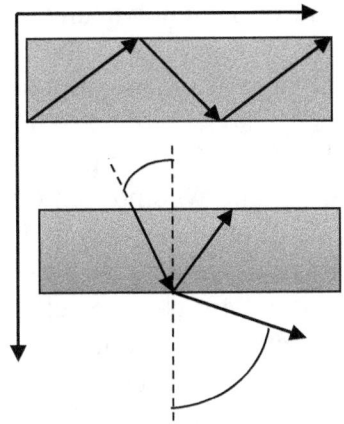

Fig.2. a) propagation of light wave in a dielectric waveguide; b) reflection and refractive of light on a boundary between free space and dielectric

light on the dielectric is equal zero. Otherwise, we have that a light wave of a finite intensity produces infinite force along a direction of propagation of light.

The same result can be obtained from known relation $f = -grad(\varepsilon)\varepsilon_0 E^2 / 2$ [Jackson 1999] for density force exerted by a dielectric located in an electrical field of strength E. Since f is proportional to the square of the electrical field, the relation can be used in an alternate electrical field of light wave. As is seen grad(ε) and f are perpendicular to the xy-plane and, therefore, the z-component of f is equal to zero. As for the x-component of the force that the pressure applied to the top plane is compensated by the force applied to the bottom plane.

Let us now consider a balance of the momentums flux densities between incident, reflected and transmitted in free space waves shown in Fig.2b. Let the energy flux of the incident wave be equal S, the energy fluxes of the reflected and transmitter waves be equal to SR and ST, respectively, where R and T are indexes of reflection and transmission. Since a lossless dielectric is considered, the following relation takes place $R+T=1$. The momentum flux of the transmitted wave in free space is know and is equal TS/c. The momentum fluxes of the incident and reflected waves in the dielectric are unknown and can be presented as $\alpha S/c$ and $\alpha R/c$, respectively where α is an unknown multiplier. From condition that a change of the z-component of the momentum fluxes is equal to zero and taking into account Fig.2b, we have $\alpha S \sin\theta_1 = \alpha SR \sin\theta_1 + ST \sin\theta_2$ or $\alpha = \sin\theta_2/\sin\theta_1$. Since in accordance with the Snell law $\sin\theta_2/\sin\theta_1 = n$ where n is the refractive index of the medium, we have $\alpha = n$ and, therefore the momentum flux in the dielectric in grater by n times than that in free space. Thus, we have derived from the Snell law and the law of conservation of energy that the momentum of light in matter corresponds to the Minkowski form. This conclusion can be considered as further evidence in favor of the Minkowski form of the momentum flux density of a continuous light wave in matter.

Doppler law

Let us compare the pressure produced on the mirror that is moving uniformly at speed v towards the light wave in the liquid dielectric as is shown in Fig. 3 and the pressure produced by the same light on the same mirror that is moving in free space. In accordance with the Doppler law, the Doppler shift of the frequency is given by $\Delta\omega = 2\omega_0[v/(c/n)]$ where ω_0 is the frequency of the light wave, n is the refractive index of the dielectric. We can see that the Doppler shift in the dielectric is greater by n times than that in free space where $n=1$. The Doppler shift is independent of the phase angle ψ_0 of the Fresnel reflection coefficient of the mirror.

Since the energy of the photon is proportional to its frequency, the energy that acquires the photons per unit time in the dielectric is greater by n times than that in free space. Since the mirror is

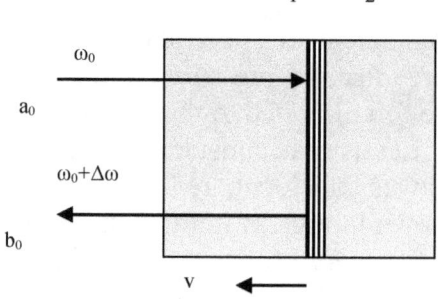

Fig.3. Designation of forward wave a_0 and backward wave b_0 reflected from the mirror 1 moving in a liquid dielectric 2 at speed v

moving uniformly in spite of the light pressure on the mirror, there should be the external pressure that compensates the light pressure. The mechanical work produced by the external pressure is equal to the increase of the energy of the photons. Then the external pressure on the mirror in the dielectric is greater by n times than that in free space. Since the external pressure is equal to the light pressure, the light pressure on a submerged mirror is greater by n times than the light pressure on the same mirror in free space. Along with other thought and real experiments, this is the next argument that the momentum of light in matter is greater by n times than that in free space.

Single split diffraction

Let a plane wave propagating in the z-direction towards a single slit in the x-y plane will undergo diffraction and produce a characteristic interference pattern in the far field. We can determine the width of the central peak of this pattern by a simple application of the Heisenberg uncertainty principle. If the slit has width Δx then the uncertainly principle requires that the field after the slit has a spread of the momentum in the x-direction of $\Delta p_x \sim h/\Delta x$. It then follows that the angular spread of the central interference peak will be

$$\theta \approx \frac{\Delta p_x}{p_z} = \frac{h}{\Delta x}\frac{c}{h\omega} = \frac{c}{\omega\Delta x}$$

If we repeat the experiment in a medium of refractive index n, then we find that the angular width of the peak is reduced by n. The momentum width Δp_x is imposed by the width of the slit, so this reduction can only arise because the momentum in the z-direction is increased by n times and, therefore

$$p_{photon} = \frac{h\omega n}{c}$$

which is the Minkowski momentum.

Evidence in favour of the Abraham form of the momentum of light in matter

A main argument in favor of the Abraham momentum in matter is the Balazs thought experiment [Balazs 1953] that was described in last time many times], [Jones 1978] , [Mansuripur 2010]. A behavior of a transparent block of an optical medium through which a light pulse is propagating is considered. The behavior is explained correctly on assumption that the Abraham momentum takes place within the block. The Minkowski momentum would predict a motion of the block in the opposite direction to the incident pulse.

The following 1D structure is considered. A light pulse of a plane light wave is propagating along the z axis in free space at speed of light c. The pulse enters a block of thickness L and refractive index $n,$. The pulse is propagating within the block, and leaves the block preserving its initial energy and momentum (Fig.4). The optical medium of the block is as simple as possible. The medium is linear, dispersionless, lossless, homogeneous, and nonmagnetic. It is supposed that measures are undertaken to exclude reflections when the light pulse enters and leaves the block. For example, the block is confined the antireflection coatings in a form of $\lambda/4$

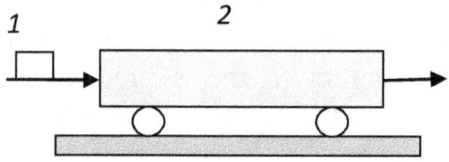

1   2

Fig. 4 Propagation of light pulse 1 through block 2 of optical medium with the reflective index n

Let us assume for the simplicity that the following condition is valid $\tau_E \ll \tau \ll T$ where τ_E, τ, and T are duration of the edges of the light pulse, duration of the pulse, and the time of propagation of the light pulse through the block, respectively (Fig.5). When the light pulse enters block 2 without reflections from free space, its speed slows from c to c/n and, as a result, it takes the time $T=nL/c$ to travel through the block. It is supposed that the block begins to move to the right when the pulse enters the block and stops its motion when the pulse exits the block. Actually, a sum of the momentums of the light pulse and block is preserved and, as a result, the momentum of the light pulse within the block decreases. When the light pulse exits the block, the momentum of the light pulse is resumed and the momentum of the block becomes equal to zero. As a result the block stops.

Let us consider two cases. First, light pulse propagates in free space outside the block. Second, the same light pulse propagates through the block. In the first case the light pulse moves in free space at the distance $cT=nL$ at time T when the light pulse in the second case moves at distance L within the block. The position of the center mass of the isolated system consisting of the light pulse and block is independent whether either the pulse is propagating through the block or the pulse is propagating outside the block. However, the center mass of the light pulse in the first case is displaced at the distance

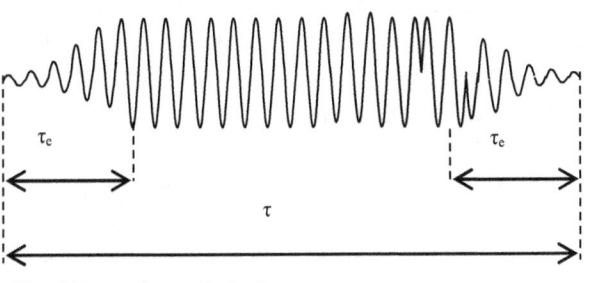

Fig. 5. Form of an optical pulse

$$\Delta Z=Ln-L=L(n-1). \tag{3}$$

In the second case, there is an additional displacement of the total center mass due to displacement of the block. This displacement is proportional to ratio of mass of the light pulse E/c^2 and mass M of the block. Here E is the energy of the light pulse. Thus, the displacement of the block is given by

$$\Delta z=\Delta Z(E/c^2)/M. \tag{4}$$

The velocity of the block is equal $v=\Delta z/T$. The momentum of the block $p=Mv=(E/c)(1-1/n)$. Since the momentum of the light pulse in free space is equal $p_0=E/c$, we have that the momentum of the block is equal to

$$p=p_0(1-1/n) \tag{5}$$

Denoting the momentum of the light pulse inside the block by x, we have from the law of conservation of the momentum $x+p_0(1-1/n)=p_0$ or $x=p_0/n$. As was noted, only the conservation of momentum and the uniform motion of the centre mass-energy in deriving this result is used and it is difficult to see how any component of this derivation could seriously be open to question.

This result has been obtained in [Barnett 2010 Phil. Trans.] on assumption that the motion of the center of the mass-energy in first and second cases should be identical. There is no notion about the mass-energy in the Newton laws. Besides, the mass of a light pulse propagating if free space at speed c in case 1 is questionable. Let us consider the similar thought experiment where these notions are not used.

Let us consider closed system shown in Fig.6 consisting of a cart 1 that can move freely along the z axis. Four plane ideal reflectors 2, 3, 4, 5 are mounted on the cart. The reflectors form a resonator of the running wave where a light wave can circulate as is shown in Fig. 1 by dashed arrows. The cross-section of the light wave S is great sufficiently so that diffraction losses can be neglected. A vessel with a transparent liquid 7 and transparent side walls 6 is located on the cart. Width of the walls is equal $\lambda/(4n^{1/2})$ where λ is the wavelength of the light radiation circulating in the resonator, n and $n^{1/2}$ are the refractive indexes of the liquid and walls, respectively. Let us assume for the sake of simplicity that the system is located in weightlessness. In this case, the pressure produced by liquid 7 on walls 6 is equal to zero. The z-component of force produced by the light radiation at a reflection from any reflector is equal to a product of a change of the momentum flux density of the light radiation W_0 in free space and area S.

Let us first consider a circulation of a light pulse shown in Fig.2. It is supposed that the momentum flux density within the pulse is equal W_0 and $\tau_e \ll \tau$ where τ and τ_e are durations of the pulse and its edges, respectively. Condition $\tau_e \ll \tau$ enables us to consider the momentum of the pulse equal to $SW_0\tau$ without taking into account the fact that the momentum flux density W within the edges of the pulse is smaller than W_0. Let the pulse be propagating and be reflecting from reflector 2. In this case a negative z-component of the momentum of the pulse equal to $-W_0\tau S$ arises. In accordance with the law of conservation of the momentum, the cart acquires positive momentum equal to $W_0\tau S$. As a result, the center mass of the cart begins to move along the z-axis at positive speed $v = W_0\tau S/M$ where M is mass of the whole system consisting of the cart, reflectors, vessel, and liquid. When the pulse is reflecting from reflector 3, its z-component of the momentum is changed from $-W_0\tau S$ to 0. As a result, the momentum of the system becomes equal to zero and the center of mass of the system stops. Let us assume further that time of propagation of the pulse between reflectors 2 and 3 $T=L/c$ is greater essentially than duration of the pulse τ. Condition $T>>\tau$ enables us to calculate simply a displacement Δz of the center of mass of the system. In this case $\Delta z=vl/c=W_0\tau Sl/(Mc)$.

When the pulse is reflecting from reflector 4, the system acquires negative velocity $v= -W_0\tau S/M$. When the pulse is reflecting from reflector 5, the velocity of the system becomes equal to zero. Time of propagation of the pulse between reflectors 4 and 5 is greater than that between reflectors 2 and 3 because the speed c/n of propagation within the optical medium at distance l is smaller than the speed c in free space. As a result, time of propagation between reflectors 4 and 5 is greater by $\Delta t=(n-1)l/c$ than that between reflectors 2 and 3. In this case we obtain that the center of mass of the system is displaced at distance $\Delta z_1=-v\Delta t=-$

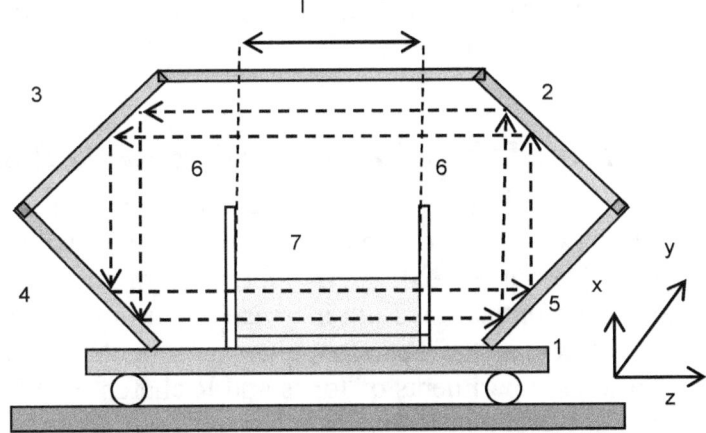

Fig. 6. A system for analysis of pressures produced by light pulse and continuous light wave in a form of a cart 7 where an optical resonator of travelling wave consisting of cart 1, reflectors 2, 3, 4, 5, vessel with transparent walls 6, and liquid dielectric 7 is located

$v(n-1)l/c<0$. This contradicts to the law of conservation of the momentum in accordance with which a motion of the center of mass of any closed system cannot be changed by inner forces. If the center of mass is dormant before the reflection of the pulse from reflector 2 in the initial state, a displacement of the center mass should be equal to zero when the system takes the initial state again.

Displacement $\Delta z_1=0$ can take place if velocity of the center of mass of the system decreases by some value Δv when the pulse is propagating within the optical medium due to the fact that a part of the momentum of the pulse is transmitted to the system. Let us calculate the part. From condition that a sum of the displacement $\Delta z_2=\Delta vnl/c$ due to the decrease of the velocity and $\Delta z_1=(n-1)l/c$ should be equal to zero we have $\Delta v/v=1-1/n$. Then the momentum p_m that acquires the system when the pulse is propagating inside the liquid can be determined from the following proportion. Since the system acquires velocity v due to the momentum $p_0=W_0\tau S$, the momentum p_m is equal $p_m=p_0\Delta v/v=p_0(1-1/n)$. Then the pressure produced by the pulse when its leading edge is entering into the medium is given by Eq.(5). This agrees with results of the known Balazs thought experiment.

Resolution of the Abraham-Minkowski dilemma

We have shown in previous sections that there are unambiguous but contradictory experiments. In accordance with thought experiments [Torchigin 2012], the momentum of a continuous light wave in an optical medium increases by n times as compared with that in free space. These thought experiments are confirmed by result of real experiments of Jones and Richards in 1954 as well as of Jones and Leslie in 1978.

In accordance with the Balazs thought experiment the momentum of a continuous light wave in an optical medium decreases by n times as compared with that in free space. These contradictory known as the Abraham-Minkowski dilemma cannot be resolved since 1910. There are several approaches in attempts to resolve the dilemma. The oldest and most widespread one is based on the energy-momentum tensor formalism. Beginning in the late

1960's something approaching a consensus emerged [Griffiths 2012]: Both the Minkowski momentum and the Abraham momentum are "correct", "but they speak to different issues, and it is largely a matter of taste which of the two (or perhaps even one of the other candidates that have from time to time been proposed) one identifies as the "true" electromagnetic momentum. Only the total stress-energy tensor carries unambiguous physical significance, and how one apportions it between an "electromagnetic" part and a "matter" part depends on context and convenience. Minkowski did it one way, Abraham another; they simply regard different portions of the total as "electromagnetic". Except in vacuum, "electromagnetic momentum" by itself is an intrinsically ambiguous notion. For example, when light passes through matter it exerts forces on the charges, setting them in motion, and delivering momentum to the medium. Since this is associated with the wave, it is not unreasonable to include some or all of it in the electromagnetic momentum, even though it is purely mechanical in nature. But figuring out exactly how and where this momentum is located can be very tricky. One would like to write down, once and for all, the complete and correct total stress-energy tensor—electromagnetic plus mechanical".

This position prevents an analysis of processes connected with transformation of the electromagnetic momentum into mechanical one and vice versa at propagation of light in an inhomogeneous optical medium and prevents to use a powerful method of a calculation a magnitude of optically induced forces (OIF) by means of an analysis of a change of the momentums. Having known a change of the momentum density fluxes in various points of the matter, the density of OIF can be calculated without knowledge a distribution of electrical and magnetic fields.

Last time several papers have been published [Mansuripur 2010, Barnett 2010 Phys.] devoted to this problem. It is declared that the problem has been resolved at last. The Barnett resolution recognizes that "the both forms are correct" and "the Abraham and Minkowski momenta are, respectively, the kinetic and canonical optical momenta". However, this approach cannot solve the problem that is formulated by Barnett as follows: "why is it that the experiments supporting one or other of these momenta give the results that they do?" Thus, this resolution gives no possibility to use the approach based on an analysis of a change of the momentum of light in practical applications.

Mansuripur believes also that he has solved the controversy. His solution is based on the generalized expression of the Lorentz force. However, this expression is incorrect because it gives incorrect result in simplest cases [Torchigin 2014 Phys. Rev.].

Various attempts to resolve the Abraham-Minkowski dilemma is reviewed in the resent paper by Kemp [Kemp 2011] where references to the Barnett and Mansuripur solutions are presented. The following conclusion is derived: "a complete picture of electrodynamics has still yet to be full interpreted".

There are the following real experiments performed by Jones *et al* [Jones 1954], [Jones 1978]. When a mirror is immersed in a dielectric liquid, the radiation pressure exerted on the mirror is proportional to the refractive index of the liquid. An accuracy of this effect was 0.05%. On assumption that the light has the momentum that changes its sign at reflection from the reflector, one can conclude that the momentum of the light in the liquid is greater by n times than the momentum of the same light in free space.

However, this conclusion contradicts to results obtained in the same time from the Balazs thought experiment [Balazs 1953] based on the generally accepted law of the momentum conservation. It is shown theoretically that a transparent block through which a light pulse is propagating without reflection should be displaced in a direction of the propagation of the light pulse. As a result, the block is moving along the light pulse when the pulse is propagating inside the block. In this case, a part of the momentum of the pulse is transferred to the block. Therefore, the momentum of light inside the block is smaller than that of the same pulse propagating in free space. Thus, the momentum of light inside the block corresponds to the Abraham form. As is pointed in review [Barnett 2010], "If argument advanced in favor of the Abraham momentum were to be incorrect, than that would bring into question uniform motion of an isolated body as expressed in the Newton's first law of motion".

Since it is impossible in a frame of the energy-momentum tensor formalism to resolve the controversy, we will consider a controversy between unambiguous results of real and thought experiments and terms connected with names of Abraham and Minkowski we will not use. In this case, momentum that is smaller by n times than that in free space we will denote as B-momentum meaning that a magnitude of the momentum is derived from the Balazs thought experiment. Accordingly, the momentum which magnitude is greater by n times in matter than that in free space we will denote by J-momentum meaning that the magnitude is derived from the Jones et al experiments. By the way, in accordance with our thought experiment [Torchigin 2012] a conclusion was derived theoretically that the momentum of a continuous light wave propagating in matter corresponds to the J-momentum. In this case, the pressure on the leading edge of the block should be given by

$$P = W_0(1-n)$$ (6)

No assumption about kinds and physical origin of OIF responsible for an increase of the momentum is made. Thus, there are two unambiguous rival thought experiments [Balazs 1953], [Torchigin 2012] where no assumption about kinds of optically induced force and their physical origin is made. These experiments are a safe ground for analysis a magnitude of momentums and OIF in matter.

Hitherto all attempts to match rival results of the experiments were failed because a reason of this discrepancy was not found out. Recognition that both results are correct because they correspond to experiment is insufficient. An explanation is required why the momentums are different. We present our explanation and show that it can be presented many years ago because the explanation is based on classical laws of mechanics.

We have to admit that the momentum flux of a light pulse in accordance with Eq. (5) decreases in matter by n times but the momentum flux of a continuous light wave in accordance with Eq. (5) increases in matter by n times. The first attempts to resolve the contradiction was undertaken by Jones [Jones 1978 365-371]. He supposed that mechanical processes should accompany a propagation of a light pulse in matter and these processes should be connected with the mechanical momentum that is moving together with the light pulse. Mansuripur assumed that the mechanical momentum is produced by the Lorentz density force arising in regions where leading and travelling edges are propagating [Mansuripur 2010]. However, his assumption is incorrect because, as is shown below, the Abraham force should be used rather than the Lorentz one.

A difference is connected with edges of the light pulse that are absent in a continuous light wave [Torchigin 2014 Optik]. Let us assume that there are additional pressures in the regions where the leading and trailing edges of the light pulse are propagating. A joint action of the additional pressure produced by the leading edge along with the negative pressure given by Eq. (6) results the positive pressure given by Eq. (5). In this case, a magnitude of this additional pressure is given by

$$P_A = W_0(n - 1/n) \qquad (7)$$

Analogously, the additional pressure produced by the trailing edge of the pulse should be equal to $-P_A$. In this case a process of propagation of a light pulse through the block in the Balazs thought experiment at $\tau < T$ looks like as follows. When only the leading edge of the pulse is propagating inside the block, there are two pressures applied to the block. Negative pressure given by Eq. (6) is applied to the front face of the block. Positive pressure given by Eq. (7) is applied to the region where the leading edge is propagating. Time instants when these pressures are terminated are identical and are equal $t = \tau$. As a result, a total pressure on the block is given by $W_0(1-n) + W_0(n-1/n) = W_0(1-1/n)$. This is in accordance with Eq.(5). Thus, the pressure applied to the block obtained from the Balazs thought experiment can be obtained on an assumption that additional pressure in accordance with Eq. (7) takes place in the region where the leading edge of the pulse is propagating. Unlike the interpretation of the Balazs thought experiment that the pressure on the front face of the block is positive and is given by Eq.(5), there is the negative pressure given by Eq.(6). Additional pressure P_A given by Eq. (7) should be taken into account to obtain the pressure in accordance with Eq. (5).

When the trailing edge enters the block, the negative pressure in accordance with Eq. (6) disappears. In the same time the negative additional pressure $-P_M$ in accordance with Eq.(7) in the region where the trailing edge is propagating arises. A sum of pressures produced by the leading and trailing edges of the pulse is equal to zero and the center of mass of the block moves uniformly.

When the leading edge leaves the block, positive pressure P_M in accordance with Eq. (6) arises on the back face of the block. In the same time positive pressure in the region where the leading edge is located disappears but the negative additional pressure produced by the trailing edge of the pulse remains. As a result, a sum of pressures is negative and is equal $-W_0(1-1/n)$. This pressure provides a negative acceleration to the center of mass of the block. The center of mass stops when the trailing edge leaves the back face of the block. This picture is in a full compliance with results of the Balazs thought experiment. As is seen no notion about the Abraham or Minkowski momentums has been used.

The density force on the region where the leading edge is propagating can be determined as follows. The energy density W of the pulse at its leading edge is changed from 0 to W_0. Then the density force is given by $f_A = dP_A/dz = (n-1/n)dW/dz$. Since $z=tc/n$, we have $f_A = \dfrac{(n^2 - 1)}{c} \dfrac{dW}{dt}$. There is no reflection at the entrance of the light pulse into the liquid. In this case the energy flux density $<S>$ inside the liquid and free space is identical and is given by $<S>=Wc$, where symbol $<>$ means an average over period of oscillation. Since $\langle S \rangle = \langle [E \times H] \rangle$, we have

$$f_A = \frac{(n^2 - 1)}{c^2} \frac{d\langle [E \times H] \rangle}{dt} . \tag{8}$$

Eq.(8) determines the Abraham force [Moller 1972], [Brevik 2009] rather than the Lorentz force. Thus, an attempt to coordinate results of unambiguous thought experiments leads to the need to recognize that the Abraham force is responsible for the pressure produced by a light pulse. An existence of this force is discussed for a long time. We have shown now that the existence of the force can be derived theoretically. It is important to underline that no preliminary assumptions about a nature and a physical origin of the optical pressure are used either for Eq.(5) or for Eq.(6) and, therefore, for Eqs. (7) and (8).

The leading edge of the light pulse propagating in an optical medium produced the pressure in accordance with (7). The pressure produces mechanical MDF in the optical medium expressed by the same (7). The trailing edge of the pulse produced the negative pressure that produced the negative mechanical MDF. As a result, the MDF takes place in the region between the leading and trailing edges of the pulse. The MDF is propagating together with the light pulse at speed c/n.

The MDF provides a displacement of the matter in the regions where it is propagating. As a result, the MDF provides a displacement of the whole block at very short time equaled to the time T of propagation of the light pulse through the block. The displacement of the block that can be calculated from the following proportion. Pressure in accordance with Eq. (5) causes a displacement of the block Δz given by Eq.(4). Then the pressure given by Eq. (7) causes the following displacement $\Delta z_1 = \Delta z(n-1/n)/(1-1/n) = \Delta z(n+1)$. The pressure produces the displacement not only the center of mass of the block but also the whole block because all parts of the block are displaced at identical distance. The negative pressure applied to the front face of the block causes the negative displacement given by $\Delta z_2 = \Delta z(1-n)/(1-1/n) = -\Delta zn$. As is seen $\Delta z_1 + \Delta z_2 = \Delta z$.

Although the negative displacement of the center mass of the block is equal Δz_2 due to the pressure given by Eq. (6), the displacement of the regions near the front face of the block is significantly greater. The displacement of the regions near the rear face of the block is equal to zero because the displacement arising near the front face is propagating at sound speed that is smaller by five orders of magnitude than the light speed. The transient processes will be terminated when the light pulse has propagated at great distance above $10^5 D$.

The additional pressures in accordance with Eq. (7) are not taken into account in the generally accepted interpretation of the Balazs thought experiment. Because of this, an erroneous conclusion that the momentum of pulse decreases in matter was derived.

As is seen, a transmission of the momentum to the block differs essentially from the simplest view accepted in the generally accepted interpretation of the Balazs thought experiment where it is supposed that the momentum of the pulse simply decreases by n times at entering the block and recovers its value at exiting the block. In reality, it is a complex procedure where the propagation is accompanied by various forces arising in various regions of space in various time instants. These forces produce various mechanical momentums of different signs in different regions of the block. The light pulse transmits to the block mechanical momentums of different signs and leaves the block at light speed. In this time, mechanical transient processes initiated by OIF are not terminated although the displacement of the center of mass of the block is equal Δz when the trailing edge of the light pulse has left the block. The inner forces that take part in the transient processes change positions of various parts of the block but do not change the displacement of the center of mass.

Thus, a joint consideration of all OIFs enables us to match rival results of the Balazs and Jones experiments. No notion about the Abraham or Minkowski momentum of light is used to interpret a behavior of the block derived from the Balazs thought experiment as well as the pressure produced on a reflector in the Jones *et al* experiments.

In a general case, a sum of forces presented by Eqs. (6) and (7) gives the following expression for OIF density produced by the light which intensity (energy density) is changed in time

$$f = -grad(n^2)\varepsilon_0 \frac{E^2}{2} + \frac{(n^2-1)}{c^2} \frac{d[E \times H]}{dt} \qquad (9)$$

Eq. (9) has long been known [Landau, 1984] and is used at present time [Brevik 2009].

A physical nature of optically induced force

We have shown that the Abraham density force is responsible for a rise of optical pressures in the regions of the optical medium where the intensity of light is changed in time. Let us now analyze

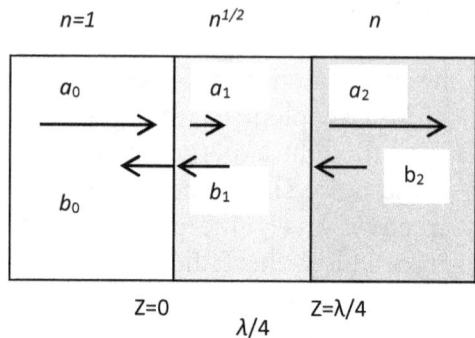

$n=1$ $n^{1/2}$ n

a_0 a_1 a_2

b_0 b_1 b_2

Z=0 $\lambda/4$ Z=$\lambda/4$

Fig.7. Propagation of a light wave from free space through antireflection plate $\lambda/4$ in the optical medium

a physical nature of force responsible for an increase of the momentum flux in an optical medium by n times. Let us consider a propagation of the light wave from free space where $n=1$ through the antireflection $\lambda/4$ plate in the optical medium of the refractive index n as is shown in Fig. 7.

Let us designate forward and backward waves in free space by a_0, b_0, respectively. Accordingly, forward and backward waves in $\lambda/4$ plate are designated by a_1 and b_1 in the left-hand boundary of the plate. Forward and backward waves in the left boundary of the optical medium are designated by a_2, b_2. It is supposed that the optical medium is infinite and, therefore, $b_2=0$. In this case $b_1=0$ also. Relations between incident and reflected waves inside $\lambda/4$ plate in Fig. 7 are the following

$$a_1 = Ta_0 - Rb_1, a_2 = Ta_1, \ b_1 = -Ra_1 \tag{10}$$

where T and R are indexes of transmission and reflection, respectively. On assumption that the square of the amplitude is equal to the energy flow, the indexes are given by

$$R = (1 - \sqrt{n})/(1 + \sqrt{n}), T = 2\sqrt[4]{n}/(1 + \sqrt{n}). \tag{11}$$

As is seen, $R^2+T^2=1$. This means that the sums of the energy fluxes before and after reflection on the boundary are identical. From Eq. (10) we have $a_1 = a_0/T$, $a_2 = a_0$, $b_1 = (R/T)a_0$, $b_0 = 0$.

The pressure produced on the boundary between free space and $\lambda/4$ plate can be determined from the following balance momentum fluxes on the boundary between free space and $\lambda/4$ plate $P = W_0[1 - \sqrt{n}(1/T^2 + R^2/T^2)]$

Here we take into account that momentum flux of light inside $\lambda/4$ plate increases by $n^{1/2}$in accordance with Eq. (6) because its refractive index is equal by $n^{1/2}$. Taking into account Eq. (11), we obtain $P = W_0(1-n)/2 = \varepsilon_0 E_0^2(n-1)/4$, where E_0 is the amplitude of the strength of the alternate electrical field on the boundary between free space and $\lambda/4$ plate. Since the refractive index inside $\lambda/4$ plate is equal to $n^{1/2}$, its permittivity $\varepsilon = n$. Then the pressure on the boundary is given by

$$P = -\varepsilon_0 E_0^2(\varepsilon - 1)/4 \qquad (12)$$

Pressure P in Eq.(12) is the pressure averaged over period of oscillations of the alternate harmonic electrical field of a light wave where E_0 is the amplitude of oscillation. In this case, the instant pressure is given by

$$P = -\varepsilon_0 E^2(\varepsilon - 1)/2 \qquad (13)$$

where E is the instant magnitude of the electrical field strength. Eq. (13) describes the pressure on the boundary where an abrupt change of the permittivity takes place. In a general case at a gradual change of the permittivity Eq.(13) is transformed as follows

$$f_{ES} = -grad(\varepsilon)\varepsilon_0 E^2/2 \qquad (14)$$

This is the first term in Eq. (9). Thus, the optical density force that increases the momentum flux of the light entering the optical medium through antireflection $\lambda/4$ plate is a counterpart of the force f_{ES} that produces pressure P that is known for a long time in electrostatics. Brevik calls this kind of force by the Abraham-Minkowski force [Brevik 2009]. Probably, it is supposed that there is an agreement between Abraham and Minkowski in an existence of this kind of force. Kemp calls this kind by the Helmholtz force [Kemp 2011]. However, this name is more suitable for internal OIF arising due to an inhomogeneity of the electrostriction pressure [Torchigin 2012]. We will call this type of force by gradient one because the force takes place in an inhomogeneous optical medium where the gradient of the refractive index is different from zero. As Maxwell was the first who studied this kind of force in electrostatics, we will call it further by the Maxwell-like force in electrodynamics.

Let us now analyze the physical nature of the OIF presented by the second term in Eq. (9). This is the Abraham force that

27

describes the time rate of redistribution between the electromagnetic momentum density of the light wave and the mechanical momentum density of the medium. Till now we considered a particular case of the optical medium where $\mu=1$ and $n^2=\varepsilon$. In a general case, we have $n^2=\varepsilon\mu$ and the second term in Eq. (9) can be rewritten in a form

$$f = \frac{(\varepsilon\mu-1)}{c^2}\{\frac{[dE \times H]}{dt} + \frac{[E \times dH]}{dt}\} = [\frac{dP}{dt} \times B] + [D \times \frac{dM}{dt}]$$
(15)

Here we designate general polarization
$$P=\varepsilon_0(\varepsilon\mu-1)E$$
(16)

and general magnetization
$$M=\mu_0(\varepsilon\mu-1)H.$$
(17)

The first term in Eq. (15) describes the density force arising at an interaction between the polarization current dP/dt and the magnetic field B. As is known, an interaction between the current of free charges in a conductor and the magnetic field is described by the Ampere law. It is not surprising that the interaction between the polarization currents and the magnetic field takes place because the polarization current is accompanied by the redistribution of dipoles in the dielectric. For example, the redistribution of the dipoles in the dielectric located between plates of a plane capacitor is accompanied by a rise of charges on the plates as if the current passes through the dielectric. This kind of force has been observed experimentally [Walker 1975, 1977] and can be called by the Ampere-like force.

Trying to keep the similarity between the first and second terms in Eq. (15), we can say that the second term describes the density force arising at an interaction between the current of magnetization and the electrical field. As is known, unlike electrons, magnetic charges are absent and the free magnetic current can not exist unlike the free electrical current. However, the magnetic dipoles exist and the magnetic alternate current due to the dipoles can be imagined. Quite recently, experimental evidence confirming an existence of the kind of forces given by the second term of Eq. (15) was published [Rikken 2012]. As is pointed out in Abstract, "The force induced by a time-varying magnetic field and a static electrical fields reported for the first time". To the best of our knowledge, the name of this kind of force is unknown. We will call it by the Ampere-twin force.

Thus, a sum of the Ampere-like and Ampere-twin forces in accordance with Eq. (15) is equal to zero at a steady state when the energy density flux that is determined by the Poynting vector $[E \times H]/c^2$ does not depend on time. The force averaged over time is equal to zero at a steady-state.

On assumption that the leading edge of the light pulse is located in region $z=0...z=l$, we obtain that the pressure produced by the leading edge can be calculated in accordance with Eq. (15) as follows

$$
\begin{aligned}
\boldsymbol{P} &= \frac{\varepsilon\mu-1}{c^2}\int_0^l (H_y \frac{\partial E_x}{\partial t} + E_x \frac{\partial H_y}{\partial t})dz \\
&= \frac{\varepsilon\mu-1}{c^2}\int_0^l (-H_y \frac{\partial H_y}{\varepsilon_0\varepsilon\partial z} - E_x \frac{\partial E_x}{\mu_0\mu dz})dz \\
&= -\frac{\varepsilon\mu-1}{2}\int_0^l (\frac{\mu_0\partial H_y^2}{\varepsilon\partial z} + \frac{\varepsilon_0\partial E_x^2}{\mu dz})dz \\
&= -\frac{\varepsilon\mu-1}{2}\int_0^l (\frac{\mu_0\partial H_y^2}{\varepsilon\partial z} + \frac{\varepsilon_0\partial E_x^2}{\mu dz})dz \\
&= \frac{\varepsilon\mu-1}{2\varepsilon\mu}(\mu_0\mu H_y^2 + \varepsilon_0\varepsilon E_x^2) \\
&= \frac{\varepsilon\mu-1}{\varepsilon\mu}(W_H + W_E) = \frac{\varepsilon\mu-1}{\varepsilon\mu}W_{TOTAL}
\end{aligned}
\tag{18}
$$

where W_{TOTAL} is the total energy density in matter.

Relation between the electrostriction pressure and optically induced forces

Studying optically induced forces we found out that generally accepted terms for designating optical forces (optically induced forces) are absent. In parallel with terms "gradient force", "electrostrictive force", "radiation pressure" we can meet terms "electrostrictive pressure", "Lorentz density force", "optical *density force*", "*optically induced force*". Besides, there are various approaches for calculation of optically induced forces.

The question arises. What types of forces should be taken into consideration in a general case? Are types listed above independent or they are different terms connected with features of a specific case? To answer this question optically induced forces in optical medium embedded in a plane optical resonator have been calculated on the base of energetic approach where no assumptions about a physical origin of these forces are done [Torchigin 2012]. It turns out that results obtained on the base of the energetic approach are identical to that obtained on the base of other approaches with one exception.

It was shown that there are three types of optically induced forces (OIF) at a steady-state situation.

1, These are OIF forces given by Eq. (14) that arise in an inhomogeneous optical medium where its permittivity is changed in space. In this case, the momentum of light is changed and, as a result, the force arises applied to the optical medium. This type of forces can be designated as gradient forces because they arise in the medium where gradient of the permittivity ε (or the refractive index $n=\varepsilon^{1/2}$) is different from zero. This kind of forces has been analyzed by Maxwell and will be called by Maxwell-like force. The same type of forces in an optical medium is responsible for the radiation pressure.

2. This is a type of force given by the first term of Eq.(15) where the generalized polarization is given by Eq. (16). We can consider this type as a generalization of the well know type that is responsible for force applied to a conductor with electrical current located in the magnetic field. This type obeys the Ampere law and can be called by the Ampere-like force. An existence of this type has been confirmed experimentally [Walker 1977] although Walker called this type incorrectly by the Abraham force.

3. This is a type of force given by the second term of Eq.(15) where the generalized magnetization is given by Eq. (17). We can consider this type as a counterpart of the Ampere-like force. Taking into account a symmetry of the Maxwell equations relatively electrical and magnetic variables in space where free charges and currents are absent, we can derive from this symmetry

that there should be the type of force that is symmetrical relatively electrical and magnetic variables. We will call this type force by the Ampere-twin force taking into account that the electrical and magnetic variables are interchanged as compared with the Ampere –like force. An existence of this type has been confirmed experimentally [Rikken 2012].

Ought to underline that these kinds of forces are produced by light and, as a result, they change the momentum of light.

Optically induced force due to an inhomogeneity of the electrostriction pressure.

An origin of the electrostriction pressure is connected with the fact that the total energy of the electric field and the energy of the dielectric located in the field decreases with the compression of the dielectric. It turns out that the energy of the electric field in the dielectric decreases faster than the elastic energy of the compressed dielectric increases The electrostriction pressure was discovered by Helmholtz in 1886 in a static electrical field. We have shown that the electrostriction pressure takes place also in a field of light waves [Torchigin 2012].

The electrostriction pressure is given by

$$P = (\varepsilon_0 \tau \frac{d\varepsilon}{d\tau} \langle E^2 \rangle / 2)$$

(19)

Here τ is the density of the dielectric, E is the amplitude of the electrical field, symbol $\langle \rangle$ means an average over period of oscillations.

The electrostriction pressure in liquids or gases is scalar. The electrostriction pressure is proportional to E^2 that can be inhomogeneous in space. As a result, there is the additional gradient force due to the electrostriction pressure and the total force density at a steady state is given by [Landau 1984].

$$f = grad(\varepsilon_0 \tau \frac{d\varepsilon}{d\tau} \left\langle \frac{E^2}{2} \right\rangle).$$

(20)

There is a generally accepted misleading conception that an inhomogeneity in the electrostriction pressure produces the gradient density force, which is also responsible for a change of the momentum of light. We have shown that this kind of OIF is produced by internal forces which net force in each elementary volume is equal to zero. As a result, this kind of OIF occurs no influence on the momentum of light [Torchigin 2013 Opt. Comm.].

For example, the net force applied to the medium in the resonator between planes $z=z_0$ and $z=z_{00}>z_0$ with regard to pressures at planes $z=z_0$ and $z=z_{00}$ is given by

$$\int \frac{d}{dz}[\varepsilon_0\tau\frac{d\varepsilon}{d\tau}\frac{E^2(z)}{2}]dz + P(z_0) - P(z_{00}) =$$
$$P(z_{00}\} - P(z_0) + P(z_0) - P(z_{00}) = 0$$

Since z_0 and z_{00} are arbitrary, a net force applied to any volume is equal to zero. Thus, an inhomogeneity of the electrostriction pressure cannot change the momentum of light.

Criticism of the approach based on the Lorentz force

Since the dilemma about a magnitude of the momentum in matter was nor solved and, therefore, it is impossible to use the approach based on an analysis of a change of the momentum, Gordon in 1973 advanced an approach based on the analysis of the density of OIF in matter. The approach is based on the belief that the Lorentz force correctly describes the OIF in matter. Since then several hundred papers have been published based on this approach. Our study of OIF properties showed that the Lorentz force approach is erroneous. Let us show that the results obtained on the Lorentz force approach contradicts to our results derived from the unambiguous experiments.

The Lorentz force density given by [Mansuripur 2010]
$$f_1 = (P \cdot \nabla)E + \dot{P} \times B \tag{21}$$

is used for calculation of the density of optically induced force (OIF). Here E and B are respectively the electric field strength and the magnetic induction associated with the light radiation, $P=(\varepsilon-1)E$ is the polarization of the optical medium. As is known, the simplest way to check any theory is to calculate results that gives the theory in the simplest cases where result are known in advance.

First, let us consider a particular static case. Let a plane light wave propagate in free space along the z axis. Let the wave incident on a plane boundary between free space and an optical medium $z=0$. Let us consider a steady-state situation where the derivatives with respect to time in Eq. (21) are equal to zero. In accordance with Eq.(21) x, y, z components of f_1 are the following

$$\varepsilon_0(\varepsilon-1)\{E_x \frac{\partial E_x}{\partial x}, E_y \frac{\partial E_y}{\partial y}, E_z \frac{\partial E_z}{\partial z}\}$$. As is seen, the direction of f_1 depends of the electrical field E.

On the other hand, since Maxwell's time, it is known that the force density in an inhomogeneous dielectric in a static electrical field is given by Eq. (14). In this case, the direction of f is independent on the direction of the electrical field. Thus, Eq.(21) gives incorrect result for this particular case.

Second, let us consider the Lorentz force in electrodynamics. Let a continuous plane light wave in which electrical field is parallel to the x-axis be propagating along the z axis in free space perpendicular to a plane boundary between free space and an optical medium at $z=0$. In accordance with Eq. (14) force f averaged over time is different from zero. On the contrary, in accordance with Eq.(21) the force applied to the boundary is equal to zero. Indeed, in this case the first term in Eq.(21) is equal to zero because $P_x>0$, $P_y=0$, $P_z=0$ and $dE/dx=0$, $dE/dy=0$, $dE/dz<>0$and, therefore, $(\boldsymbol{P}\cdot\nabla)\boldsymbol{E}=0$. The second term in Eq.(21) is also equal to zero because the phase shift in a plane wave between multipliers $d\boldsymbol{P}/dt$ and $\boldsymbol{B}$ is equal to $\pi/2$. Averaging over time their production gives zero. Thus, in accordance with Eq. (21) the pressure produced by a plane light wave incident normally from free space on a plane surface of an optical medium is equal to zero. Thus, in this case Eq. (21) is incorrect again.

Third, as was shown [Torchigin 2014 Phys. Rev.], the Lorentz force density is different from zero in a homogeneous optical medium inside $\lambda/4$ plate. This is error. As is known, the momentum of light wave propagating in a homogeneous optical medium is constant within the optical medium. The same is valid for any number of light waves. Thus, there is no change of the momentum of light in a homogeneous optical medium. In this case, the following question arises. In accordance with the third Newton law, there should be a counterpart of any the Lorentz force f. The counterpart is applied to another object, is equal to f by value and is directed opposite f. What is a counterpart of the Lorentz force in a homogeneous optical medium? The counterpart can be applied to the light wave only. In this case, it should change its momentum. Such waves are absent in a homogeneous optical medium. Since the counterpart is absent, the OIF should be absent also.

This conclusion requires clarification. The following conclusion is more correct. Since the counterpart is absent, a net force of all kinds of OIF is equal to zero. As we have shown the force in a homogeneous optical medium imbedded in a field of a light wave is presented by Eq.(9). This is the Abraham force that describes the time rate of redistribution between the electromagnetic momentum density of the light wave and the mechanical momentum density of the medium. The force averaged over time is equal to zero at a steady-state when the energy flux density is not changed in time.

Let us compare results of calculation of OIF in matter in particular cases on the basis of our notion and notions based on the Lorentz force approach used by Mansuripur [Mansuripur 2010 Opt. Comm.]. Mansuripur studies an exchange between momentums of light pulse $\mathcal{E}_{pulse}/c$ in free space that enters the block with the refractive index n through anti-reflection $\lambda/4$ plate. Here $\mathcal{E}_{pulse}$ is the energy of the light pulse. He believes that the mechanic and electromagnetic components of the light pulse are equal respectively $(n-1/n)\mathcal{E}_{pulse}/(2c)$ and $\mathcal{E}_{pulse}/(nc)$. The total momentum of the pulse is equal to their sum $(n+1/n)\mathcal{E}_{pulse}/(2c)$. Besides, the mechanical momentum $-(n-1)^2\mathcal{E}_{pulse}/(2nc)$ is transmitted to $\lambda/4$ plate when the pulse enters the block. A sum of this momentum and the total momentum propagating in the block is equal to the momentum of the pulse in free space $\mathcal{E}_{pulse}/c$. He rightly notes that the law of the conservation of the momentum holds. However, since MDF of the pulse in accordance with Mansuripur is equal to $W_0(n+1/n)/2$, the pressure on the reflector is not proportional to the refractive index n that contradicts to results of the Jones experiment.

Our total MDF in matter is equal to $W_0 n$ and, therefore, the pressure on the reflector is proportional to the refractive index n that agrees with results of the Jones experiment.

Differences between Mansuripur and our analysis are the following. Our total momentum of the pulse, unlike $(n+1/n)\mathcal{E}_{pulse}/(2c)$, is equal to $n\mathcal{E}_{pulse}/c$. Our mechanical momentum, unlike $(n-1/n)\mathcal{E}_{pulse}/(2c)$ is equal $(n-1/n)\mathcal{E}_{pulse}/c$. At last, our mechanical momentum transmitted to $\lambda/4$ plate, unlike $-(n-1)^2\mathcal{E}_{pulse}/(nc)$ is equal $(1-n)\mathcal{E}_{pulse}/c$. As is seen, a magnitude of the mechanical component of MDF in accordance with Mansuripur $W_0(n-1/n)/2$ is half of our mechanical MDF given by Eq. (7).

To the best of our knowledge, Jones in 1978] was the first who concluded that the total momentum of light pulse in matter is equal pn, the electromagnetic momentum is equal to p/n and the mechanical momentum is $[1-1/n^2]$ part of the total momentum pn, where p is the momentum of the light pulse in free space. However, he cannot explain a physical phenomenon responsible for a rise of the mechanical momentum. He wrote "We are not able to specify the details how this body impulse is created, but merely point out it is demanded by the simple consideration of mechanics".

Mansuripur in 2010 was the second who showed that the mechanical momentums are created by leading and trailing edges of the light pulse. However, he used an approach based on the Lorentz density force and his magnitude of the total momentum of the light pulse in matter contradicts to result of Jones *et al* experiment.

We are third who have disclosed a physical origin the Jones "bodily impulse", corrected calculation of Mansuripur and showed that the A-force is responsible for a rise of the mechanical momentum in matter rather than the Lorentz force. The main difficulty was to persuade that an approach based on the Lorentz force is inconsistent in spite of the fact that it is used everywhere in last 40 years.

In summary, we have shown that the approach based on the Lorentz force given by Eq. (21) that describes the optically induced force in an optical medium and that is used last 40 years in many publications related to optically induced force is erroneous.

Resolution of debates on the She et al experiment

Seemingly, theoretical grounds presented by She *et al* in their paper are justified and corresponds to the classical notion of the momentum, which change entails a rise of an optical force. If the momentum of the light radiated from the fiber is changed, there should be forces responsible for this change. It does not matter how these forces are called. Brevik calls them by the Abraham-Minkowski force. Mansuripur supposes that these forces are called by the Lorentz force. She *et al* are interesting whether the momentum of light radiated from the fiber is greater or smaller than that within the fiber. They are not interesting in their experiment the reasons of a change of the momentum. They consider that any behavior of the nano-fiber is explained by an action of the external force arising due to a change of the momentum of light.

On the basis of our notions, the force applied to the face of the fiber in accordance with the Eq. (4) is directed towards free space and, therefore, unlike the She et al conclusion, OIF is the pull force that stretches the fiber.

Brevik asks in his comment [Brevik 2009] "But does this experiment measure electromagnetic momentum? In our opinion the answer is no. What is detected is merely the electromagnetic Abraham- Minkowski force density $f^{AM} = -(\varepsilon_0 / 2)E^2 \nabla n^2$ in the surface layer of the filament (or in other regions where n varies). This is not related with the electromagnetic momentum in itself. The electromagnetic force density is

$$f = f^{AM} + [(n^2 - 1)/c^2]\partial / \partial t(\boldsymbol{E} \times \boldsymbol{H}) \tag{22}$$

and electromagnetic momentum does not appear until the second term in this expression".

We have shown that the second term is equal to zero at a steady-state. Certainly, there is the steady-state situation in the experiment because the time constant of transient processes is smaller one picoseconds. A duration of light pulses in the experiment is greater by nine orders of magnitude. In this case, only the first term in Eq. (18) should be analyzed. This term is identical to Eq. (12) that the Maxwell-like OIF describes.

This is the simplest example that illustrates that f^{AM} can be calculated on the basis of the CM approach. Let us calculate OIF exerted on a plane boundary of an optical medium at normal incidence of a light wave from free space. The wave propagates from left to right. The momentum flux density of the wave is generally accepted for free space and is equal to W_0, where $W_0 = \varepsilon_0 E_0^2/2$ is the density of the energy in vacuum. Indexes of reflection and transmission are given by $R = (\dfrac{1-n}{1+n})^2$, $T = \dfrac{4n}{(1+n)^2}$, respectively. Let us consider the momentum of light propagating thought unit area per one second. This momentum is equal to W_0. A part of this momentum is reflected. This part is equal to $R = (\dfrac{1-n}{1+n})^2$. Another part penetrates into the medium. This part is equal to $\dfrac{4n}{(1+n)^2}$. However, the momentum in the medium increases by n times and, therefore, becomes equal to $W_0 \dfrac{4n^2}{(1+n)^2}$. Taking into account that a change of the momentum at reflection from the boundary is equal to a difference between the input and output MDFs, we have $\Delta p = p_0|p_0| + |p_R| - p_T$, where p_T, p_R and p_0 are momentums of transmitted, reflected and incident waves, respectively, we have

$$\Delta p = W_0[1 + (\frac{1-n}{1+n})^2 - 4\frac{n}{(1+n)^2} = -2W_0\frac{n-1}{n+1} \tag{23}$$

The change of the momentum is negative. Therefore, the momentum of the light in free space is smaller than that in the fiber. The force applied to the light is directed opposite the propagation of the light. The counterpart of this force is applied to the matter and directed in parallel to the propagation of the light. Thus, the counterpart is applied to the fiber and is directed along the propagation of the light.

In fairness, it should be noted that there are other grounds that that overcome above analyzed OIF. Let us list objections that give grounds to reject a conclusion derived by She *et al* that they observed OIF in accordance with the Abraham momentum of light. Moreover, there is no ground to conclude that they observed a compression of the nano-fiber.

Other inner forces produced by the light within the nano-fiber ought to be taken into account. A net force of inner forces is equal to zero and they do not change the momentum of light. The simplest example is the electrostriction pressure.

The electrostriction pressure in a homogeneous optical medium compresses the medium. As a result, the length of the fiber is decreased. In accordance with the Lorentz-Lorenz relations we have the following pressure produced by the electrostriction pressure

$$\frac{\tau}{\varepsilon}\frac{d\varepsilon}{d\tau}\langle W\rangle = \frac{1}{6}(\varepsilon-1)(\varepsilon+2)\langle W\rangle \cong 0.51\langle W\rangle.$$

On the other hand, the pressure on the face of the nano-fiber as

$$\frac{1}{n}\frac{n-1}{n+1}\langle W\rangle \cong 0.12\langle W\rangle$$

is shown in [She 2008] is given by. Thus, the electrostriction pressure that compresses the nano-fiber is greater by 4 times than a potential pull pressure that, as it is expected, should expand (or compress) the nano-fiber due to a change of the momentum of light. Thus, the nano-fiber is exerted by push force independently on whether Minkowski or Abraham momentum of light in an optical medium. Besides, there are surface OIF in accordance with the first term in Eq.(9) that tends to expand the nano-fiber in a transverse direction. There are surface OIF in accordance with second term in Eq. (20) that tends to compress the fiber in a transverse direction.

2. It is much more important that there can be inner forces that are not connected at all with the optically induced forces. For example, the observed motion can be connected with a change of inner mechanical tensions inside the fiber in its various regions. The power of light is sufficient to heat up the nano-fiber which volume is extremely small. Indeed, for given parameters of the nano-fiber $L=1.5$ mm, $d=450$ nm, $\rho=2.2 \ 10^3$ kg/m^3 the mass of the fiber $m=5.25 \ 10^{-16}$ kg is extremely small. A change of the

temperature of a body is given by $\Delta t = E/(Cm)$ where E is the energy transmitted to the body, C is the heat capacity of a glass, m is the mass of the body. Since the $C=0.84$ kg/(kJ K), the energy $E=2.14 \ 10^{-14}$J is sufficient to increase the fiber temperature by 100 C^0. This energy is smaller by 12 orders of magnitude than the energy that enters the fiber per one second at laser power $P=19.5$ W. As is pointed by She et al the fiber can be heat up at least to the temperature 273 C^0. Such heating is sufficient to change inner tensions. As a result, an initial balance of tensions is violated and fiber is bent due to the imbalance. When the light is turned off, the fiber returns in its initial position in several seconds [She 2008] due to a slow restoration of the previous tensions.

3. There is the one more proof that the observed motion of the fiber is not connected with a change of the momentum of light. The nano-fiber is a mechanical system and its behavior at an instant rise of OIF is predicted. Any mechanical system is characterized by inherent time constants that are independent on forces applied to the system. The nano-fiber is an oscillating system that is characterized by period of oscillation T and quality factor Q. The period of oscillation of the oscillation system in a form of elastic rod which one end is fixed is given by [Landau 1976 Mechanics]

$$T = \frac{16L^2}{\pi}(\rho/Yd^2)^{1/2}$$

(24)

where ρ and Y are the density and the Young's elastic modulus of glass, respectively. For given in [She 2008] parameters $L=2.5$ mm, $d=450$ nm, we have $T=25$ms. Quality factor Q of the glass system is significantly greater than 1. An instant rise of any constant OIF should be accompanied by an oscillating transient process with period T about 25 ms and duration about TQ. These processes are not observed.

4. As follows from theoretical considerations presented by She et al, the observed effect should be proportional to the power in the nano-fiber. The greater power enters the fiber the greater the observed effects. However, observations with the light power that is essentially greater than 19.5 mW are not presented. If this dependence were observed, it were presented.

5. Observed movement of the nano-fiber in a horizontal direction at distance about 15 μm in several seconds can be explained only by physical phenomena which time constant is essentially greater. In this case, the observed motion of the nano-fiber is not connected with OIF.

6. Ought to draw attention that a use of the nano-fiber for an analysis of OIF applied to its face is not justified. Actually, with a gradual decrease of fiber diameter *a*, a part of light energy propagating inside the fiber is gradually decreasing and tends to zero when *a* tends to zero [Govar 1989]. A total light energy that is transmitted by the fiber is unchanged. As a result, a part of the energy propagating in vacuum around the fiber as well as the area around the fiber where the energy of the light wave takes place are gradually increasing with *a*-> 0. The light wave directed by the fiber tends to a plane light wave of TE type. A direction of propagation of this wave coincides with the fiber axis. Ought to underline that there are no diffraction expansion of the light wave until fiber exists.

If the fiber is interrupted abruptly and there is a face of the fiber of small diameter at $z=0$, a main part of the light wave propagating from $z=-\infty$ to $z=0$ around the fiber in free space continues to propagate from $z=0$ to $z=\infty$. Only a small part of the light wave energy propagating within the fiber is reflected. As a result, the Fresnel formulas used by the She *et al* that are valid for plane waves incident on a plane boundary occur invalid for analysis of reflections from the nano-fiber face. A purpose of the She *et al* experiment was to observe a maximal effect connected with reflection the light wave propagating in an optical medium from the fiber face. This effect is significantly smaller for nano-fiber than for a conventional fiber. A making use of the nano-fiber is not reasonable.

As for the Mansuripur comments, they are based on the Lorentz force approach. We have shown that the approach is inconsistent and erroneous. There are many other misconceptions in the debated. However, in our opinion there is no necessity to list all of them to conclude that at present there is no generally accepted understanding about kind of OIF and momentums of light. Because of these, it is naïve to hope that the optical theory of ball lightning based on the optically induced force can win a general acceptance in the nearest future. At least, a new generation of scientists is required.

We can conclude that a single thing that showed the She et al experiment is the fact that a notions about optically induced forces are absent completely in even the simplest set up. Because of this it is naïve to wait that the optical theory of Ball Lightning where optically induced forces play a decisive role will be accepted.

Conclusion

On the basis of physical laws that require no assumption about kinds of optically induced forces and the quantity of these kind, we have shown that the momentum of light in matter is greater by n times than that of the same light in free space. Here n is the refractive index of the matter. If the momentum of a light pulse is considered, that it consists of two components. This is the electromagnetic one that is equal to $1/n^2$ part of the total momentum and the mechanical component that is equal to $(1-1/n^2)$ part of the total momentum.

An alternative treatment for the momentum of a light pulse is possible. A propagation of a light pulse the momentum of which is greater by n times than that of the same light pulse in free space is accompanied by two additional optically induced forces applied to the regions of the matter where the leading and trailing edges of the light pulse are propagating. These forces transmit a part of the total momentum into the mechanical component that is located between the edges and id propagating together with the light pulse at speed c/n. The mechanical component is connected with the mechanical motion of the matter.

An origin of the optically induced forces arising at a change of the momentum density flux of a continuous light wave is explained by the Maxwell-like kind of force that acts on a dielectric located in an electrical field.

An origin of the force arising at propagation of a light pulse in a homogeneous optical medium is explained by action of two different kinds of forces. The first kind can be called by the Ampere-like force that acts on a conductor with electrical current located in a magnetic field. Although a current of free charges is absent in a dielectric, a change of the polarization of the dielectric in time can be considered as the same current.

The name of the second kind is absent till present day. This kind is dual to the Ampere-like kind. The force arises at an interaction between the "magnetic current" due to the alternate magnetization and an electrical field. Magnetic free charges are absent in the nature and, therefore, the kind of force that is dual to the Ampere force does not exist. However, magnetic dipoles are presented in the nature and, therefore, the kind of force that is dual to the Ampere-like force does exist.

At a steady-state when the intensity of light is not changed in time, a sum of the Ampere-like and Ampere-dual forces averaged over a period of oscillations is equal to zero

There is a perfectly different kind of optically induced force arising at compression of the optical medium due to the electrostriction pressure. A magnitude of this kind of force is equal to the gradient of the electrostriction pressure and is comparable with the Maxwell-like force. However, forces of this kind are produced by inner forces rather than a change of the momentum of light. Thus, this kind of force can not be calculated by means of an analysis of a change of the momentum of light.

On the basis of presented notions about optically induced forces, the Abraham-Minkowski controversy is absent because the initial notions about the momentum of light inmate are modified. In the same time it is shown that that the approach to calculation of optically induced forces on the basis of the Lorentz force that is considered at present as a main instrument for calculation of optically induced force is erroneous.

The presented theory enables us to analyze debates about explanation of results of the recent experiment regarding measurement of the momentum of light in matter. We have shown that a generally accepted notions about kinds of optically induced force optically induced is absent at present. We hope that our book will eliminate this drawback.

References

Abraham, M. R. C. Circ. Mat. Palermo 28, 1 (1909); 30, 33 (1910).
Balazs N. L., Phys. Rev. 91, 408-411 (1953).

Barnett S. M. and R. Loudon Phil. Trans. R. Soc.A 368 927 (2010).

Barnett S.M., Phys. Rev. Lett. 104, 070401 (2010).

Brevik I. Phys. Rev. Lett. 103, 219301 (2009).

Burt M. G. and R. Peierls, Proc. Roy. Soc. London, Ser. A 333, 149-156 (1973).

Griffiths D.J. Am. J. Phys. 80 7 (2012).

Jackson J. D. Classical electrodynamics, (Wiley, N.Y., 1999).

Jones R. V. and B. Leslie, Proc. Roy. Soc. London, Series A, 360, 347-363 (1978).

Jones R. V. and J. C. S. Richard Proc. R. Soc. London Ser. A 221, 480 (1954).

Jones R. V., Radiation pressure of light in a dispersive medium, Proc. R. Soc. Lond. A 360 (1978) 365-371.

Kemp B. A., J. A. Kong, and T. Grezegorczyk, Phys. Rev. A 75, 053810 (2007).

Kemp B. A., Journal of Applied Physics 109 111101 (2011).

Kemp B., T. Grzegorczyk, and J. Kong, Opt. Express 13, 9280-9291 (2005).

Landau L. D. , E. M. Lifshits, Mechanics. Vol.1 (Buttrworth Heinemann 1976).

Landau L. D., E. M. Lifshits and L. P. Pitaevski, Electrodynaics of Continuous Media (Oxford, Heinemann, 1984).

Mansuripur M. and A. R. Zakharian Phys. Rev. A 80, 023823 (2009)

Mansuripur M., Opt. Comm. 283 1997 (2010)

Mansuripur M., Opt. Comm. 283 3557 (2010)

Minkowski H., Nachr. Ges.Wiss.Gottingen 53 (1908); Math.Annalon 68, 472 (1910).

Moller R. N. C., The theory of Relavity, 2nd ed. (Clarendon Press, Oxford, 1972),

Rikken C.L.J.A., B.A. Tillelen, Physics Review Letters 108 (2012) 230402.

She W., J.Yu, and R Feng, Phys.Rev. Lett., 101 243601 (2008).

She W., J.Yu, R. Feng, Phys. Rev. Lett. 103, 219302 (2009).

Torchigin V.P., Torchigin A.V. Comment on ``Transverse radiation force in a tailored optical fiber" Physical Review A 2013 Vol. 88 p. 027801 013Torchigin V.P., Torchigin A.V. Compensation of the optically induced Lorentz force in a homogeneous optical medium Optik 2013 Vol.124, p.5492-5495

Torchigin V.P., Torchigin A.V. Optically induced force in a curve lightguide. EPJ AP European physical journal Applied Physics 2013, vol. 63, p. 10501

Torchigin V.P., Torchigin A.V. Comment on "Theoretical analysis of the force on the end face of a nanofilament exerted by an outgoing light pulse" Physical Review A 2014, vol. 89, page 057801

Torchigin V.P., Torchigin A.V. Comparison of various approaches to the calculation of optically induced forces, Annals of Physics,Volume 327, Issue 9, September 2012, Pages 2288–2300 2012

Torchigin V.P., Torchigin A.V. Interrelation between striction forces in dielectrics and optically induced forces in transparent media Physica Scripta Volume 86 Number 2 2012 86 025402

Torchigin V.P., Torchigin A.V. Magnitude of the photon momentum in matter. American Journal of Science and Technology 2014 Vol.4 Nom. 4 page 151-156

Torchigin V.P., Torchigin A.V. Pressure Exerted on a Semi-Infinite Lossles Dispersionless Dielectric by a Plane Electromagnetic Wave OPEN JOURNAL OF MODERN PHYSICS 2014 Vol. nom. 3 1 page 1-7.

Torchigin V.P., Torchigin A.V. Propagation of a light pulse inside matter in a context of the Abraham–Minkowski dilemma. Optik 2014 vol. 125, issue 11, pp. 2687-2691.

Torchigin V.P., Torchigin A.V. Resolution of the Age-Old Dilemma about a Magnitude of the Momentum of Light in Matter Physics Research International 2014, Vol. 2014, Pages 126436.

Torchigin V.P., Torchigin A.V. The momentum of an electromagnetic wave inside a dielectric derived from the Snell refractive law Annals of Physics 2014, vol. 351, pages 444-446

Walker G. B., D. G. Lahoz, and G. Walker, Can. J. Phys. 53, 2577 (1975).

Walker G.B., G Walker Nature 265 324 (1977).

Walker G.B., G Walker, Can. J. Phys. 55 2121 (1977).

Appendix

Comparison of various approaches to calculation of optically induced forces

V. P. Torchigin, A.V. Torchigin

Institute of Informatics Problems, Russian Academy of Sciences, Nakhimovsky prospect, 36/1, Moscow, 119278, Russia, tel +7 499 1332532, email: v_torchigin@mail.ru

PACS: 42.65.Jx, 42.65.Tg

Keywords: optical electrostriction; optical forces; Maxwell stress tensor; Lorentz density force; Helmholtz pressure.

Abstract

Various approaches used for calculation of optically induced forces applied to a transparent optical medium imbedded in a closed plane optical resonator are analyzed. The forces are calculated by means of analysis of a change in the eigen frequency and energy stored in the resonator at various positions of the medium. It is shown that results obtained are identical to that calculated by means of approaches based on the Maxwell stress tensor, based on an analysis of a change in the momentum of light. An exception is for results obtained on the base of last versions of the Lorentz density force.

1. Introduction.

At present, there are five at least approaches to calculation of optically induced forces. The first one supposes that results obtained in electrostatics [1-9] can be used for calculation of optically induced forces (ES approach). This assumption is based on the fact that forces in electrostatics are proportional to the square of the strength of electrical field E^2. In this case the square of the strength of alternate electrical field of light wave averaged over cycle is different from zero and can be used instead of the E^2. There are no mentions about the magnetic field of light waves.

45

This approach assumes that natures of forces in electrostatics and electrodynamics are identical. There are two types of optically induced forces in electrostatics. These are the Maxwell gradient density force (force per unit volume) arising in an inhomogeneous dielectric and the Helmholtz electrostriction pressure arising in any dielectric owning a change in its permittivity with a change in its density. The Maxwell-like force in optics is called usually the radiation force [10] or the radiation pressure [11-14]. The Helmholtz-like pressure in optics is called usually the optical electrostriction pressure [8] or electrostrictive forces [15].

The second approach is based on the *Maxwell stress tensor* (MST). It is supposed that the knowledge of the full electromagnetic field distribution in an optical system is a prerequisite for the computation of optically induced forces. In these electromagnetic field-based calculations, MST is numerically integrated over a closed surface surrounding the medium to compute the optical forces acting on it [15]. In this case the magnetic field of light waves is taken into account. A term *"radiation pressure"* is used for designation of this type of forces. Recently in 2010, MST approach was supplemented by the electrostrictive components of optical forces [15]. A term *"electrostrictive forces"* was introduced along with term *"radiation pressure"*. It is supposed that, just like MST, *the electrostrictive force* can be found from *the electrostriction pressure* distribution. This approach is used at present in nano-optics [16].

The third approach is based on an analysis of a change in the momentum of light (CML) in a transparent optical medium. It is supposed that any reflection of light wave is connected with a change in its momentum of light. This means that there is a force applied to the light. Accordingly to the third Newton law, there is a force applied to the optical medium. The force is equal to force applied to the light by value but is opposite by direction. However, at present there is no general accepted notion about a magnitude of the momentum of light in an optical medium. Minkowski and Abraham proposed two different approaches to this problem. A great deal of controversy has centered on the rival claims of the Abraham expression for the photon momentum $h\nu/n$ where n is the refractive index of the medium. The Minkowski momentum is greater by n^2 times. The few available experiments seem to favor the latter, while many theories favor the former. This is not the place to review the large literature devoted to this problem. The interested reader can study the recent review by Pfeiffer et al. [17]. In 2010 various arguments in favor of Abraham and Minkowski theories were presented in review [18] devoted to the 100-year anniversary of this problem.

The fourth approach is connected with an idea of Gordon [10]. He proposed to calculate radiation forces applied to the optical medium by using the Lorentz law for density force. Having known a distribution of radiation forces in space, one can determine a distribution of the flow of the momentum in space. Since then several tens of papers has been published where the force of the electromagnetic radiation on material objects is derived by a direct application of the Lorentz law of classical electrodynamics. For example, we can mention papers of Loudon, Barnett, Mansuripur, Kemp, Obukhov, Leonhardt, Zakharian, Maloney, Boyd. There are no mentions about a difference between the radiation forces and optical electrostriction pressures. A single term "radiation pressure" is used usually to denote optically induced forces. Probably, it is supposed that a study of the problem in a generalized form enables to take into account all possible types of optically induced forces. At present, it is known two different forms of the Lorentz force density, at least, proposed by Gordon [10] and Mansuripur [13]. It is hard to agree that both forms are correct. Possibly, one or two of them are incorrect.

At first sight, it is difficult to believe that the ES approach, where notions of static electric field and permittivity of a dielectric are used can give the same results as CML or Gordon approaches where perfectly different notions such as light waves, magnetic field, radiation, reflective index, momentum of light are used.

There is the fifth safe approach to solution of this problem. Following works of Rakich, Povinelli et al. [14, 19-21], optical forces applied to an optical medium are calculated by a perfectly independent way by means of analysis of a change in the eigen frequency of an isolated closed oscillating system in a form of an optical resonator. We will use this approach for calculation of optically induced true forces because there are no previous assumptions and well-known facts like the law of conservation of energy and properties of optical resonators are used. For this purpose we place an optical medium in the simplest one-dimensional structure in a form of a plane optical resonator and calculate changes in the eigen frequency of the resonator at various positions of the optical medium. The frequency of electromagnetic oscillations in a closed oscillating system and the energy stored within it are changed simultaneously in a direct proportion with a slow change in parameters of the system. Having known a change in the eigen frequency, we can determine a change in the energy and, therefore, the net forces applied to the optical medium in each its position. These forces can be compared with the forces obtained from various approaches and the most suitable approach can be chosen. Besides, it is possible to determine what types of forces are independent and what types of forces are differed on their names only.

A methodical character of the paper induces us not to aspire to brevity, not to avoid to present formulas and expressions well-known in electrodynamics and to consider the simplest situations to minimize mathematics.

2. An energetic approach to optically induced forces applied to a transparent optical medium in a plane optical resonator

The simplest situation is considered where an optical medium is imbedded in one-dimensional oscillating structure in a form of plane optical resonator consisting of two plane parallel ideal reflectors, between which a plane wave is reflected in serial. The medium is as simple as possible. It is linear, lossless, motionless, non-dispersion, non-magnetic, homogeneous, isotropic, incompressible, without free charges liquid. Incompressibility is required to neglect the elastic energy stored in the optical medium. A steady state is considered. In this case relations between amplitudes and phases in all points of the medium are independent on time. This situation is typical for works devoted to optically induced forces.

On assumption that the z axis is perpendicular to reflectors, a field of a plane standing wave is given by

$$E_x = 2E_0 Sin(knz)Cos(\omega t), E_y = 0, E_z = 0, H_x = 0,$$
$$H_y = 2H_0 Cos(knz)Sin(\omega t), H_z = 0$$

$$(1)$$

where E_0 and H_0 are the amplitudes of the strength of electrical and magnetic fields, respectively, n is the reflective index of the optical medium, $k=\omega/c$, ω is angular frequency of the light wave, $H_0=nE_0/Z_0$, $Z_0=(\mu_0/\varepsilon_0)^{1/2}$. The length of the resonator is equal to $L=(\lambda/2)N$, where $\lambda=2\pi/(kn)$ is the wavelength of the light wave, N is integer. Unlike the constant electric field strength E_z in a plane capacitor, the electric field strength E_x and magnetic field strength H_y in the plane resonator are changed in time and space (along the z axis). Undoubtedly, the density W_{EH} of total energy stored in the resonator depends not only on the electrical field but also on magnetic field. The density of the energy (both electrical and magnetic) of a travelling light wave propagating in a homogeneous optical medium is given by

$$W_{EH} = \varepsilon_0 \varepsilon E_0^2 / 2 = \varepsilon_0 n^2 E_0^2 / 2$$

$$(2)$$

where ε and n are the permittivity and reflective index of the medium. The time-averaged density of the energy of a standing light wave in the plane resonator expressed by eq.(1) is given by

$$W_{EH} = 4\varepsilon_0 \varepsilon E_0^2 Sin^2(knz) / 4 + 4\mu_0 \mu H_0^2 Cos^2(knz) / 4 =$$
$$\varepsilon_0 n^2 E_0^2 = \mu_0 \mu H_0^2$$

$$(3)$$

As is seen, the density W_{EH} is independent on z and is greater by two times than that in eq.(2) because there are two travelling waves of identical amplitudes propagating in opposite directions. The total density is equal to a sum of densities of the travelling waves. The same is valid for travelling waves which amplitudes are different. As is seen from Eq.(3), there are time instants at which the energy stored in the resonator is presented completely in an electric form and other time instants at which the energy stored in the resonator is presented completely in a magnetic form. For example, the energy is concentrated in the electrical field at $t=0$ and the energy is concentrated in the magnetic field $t=\pi/(2\omega)$.

2.1 Optical forces applied to a sensor confined by $\lambda/4$ plates embedded in a plane resonator

Let a part of the medium in a plane resonator with the permittivity ε_1 and reflective index $n_1=\varepsilon_1^{1/2}$ be substituted by a sensor in a form of a region of permittivity ε_2 and reflective index $n_2=\varepsilon_2^{1/2}$. The sensor is confined by $\lambda/4$ plates with the permittivity $\varepsilon_{12}=(\varepsilon_1\varepsilon_2)^{1/2}$ and reflective index $n_{12}=(n_1n_2)^{1/2}$ as is shown in figure 1a. It is known that $\lambda/4$ plates are used in optics to exclude reflections from a boundary of two optical mediums with different refractive indexes. Thus, there are no reflections from $\lambda/4$ plates. Let there be a resonance in the resonator with the inserted sensor and, therefore, the phase shift of light wave at bypassing the resonator is equal $2\pi N$, where N is integer. Since there are no reflections from the sensor, the phase shift is not changed with moving the sensor along the z axis. As a result, conditions of the resonance are preserved and the eigen frequency of the resonator is not changed. In this case the energy Σ stored in the resonator is not changed also and, therefore, a net optical force applied to the sensor is equal to zero. This conclusion is valid for the sensor with arbitrary width w. There are grounds to suppose that density forces applied along width w of the sensor are absent because a net force exerted on the sensor is independent on its width w. To dispel possible doubts relatively a change of forces applied to $\lambda/4$ plates with a change in width w, let us consider forces applied to $\lambda/4$ plate.

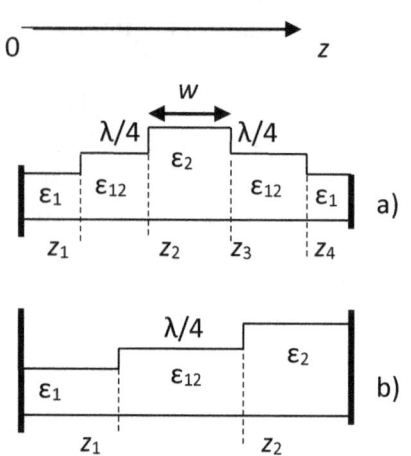

Figure 1. Position of $\lambda/4$ plate (a) and sensor with two $\lambda/4$ plates (b) in a plane optical resonator

2.2 Forces applied to λ/4 plate in a plane optical resonator

Let a part of the medium in a plane resonator with the permittivity ε_1 and reflective index $n_1 = \varepsilon_1^{1/2}$ be substituted by a sensor in a form of a region of permittivity ε_2 and reflective index $n_2 = \varepsilon_2^{1/2}$ confined by $\lambda/4$ plate of permittivity $\varepsilon_{12} = (\varepsilon_1\varepsilon_2)^{1/2}$ and reflective index $n_{12} = (n_1 n_2)^{1/2}$ -as is shown in figure 1b. Since there are no reflections from $\lambda/4$ plate, the phase shift at bypassing of the resonator by a wave in a form $\mathrm{Exp}[i(\omega t - knz)]$ is given by

$$\varphi = -2k[n_1 z_1 + n_{12}\lambda_{12}/4 + n_2(L - z_1 - \lambda_{12}/4)] = \\ -2\pi N_1 \tag{4}$$

where $k = \omega/c$ is the propagation constant, ω is the light frequency, L is the distance between reflectors, $\lambda_{12} = \pi/(2n_{12}k)$ is the width of $\lambda/4$ plate, N_1 is integer.

The eigen frequency of the resonator is changed with a change in L in such a way that eq. (4) remains valid. In this case eq.(4) can be considered as an implicit function between k and L and

$$\frac{dk}{k} + \frac{n_2 dL}{n_1 z_1 + n_{12}\lambda_{12}/4 + n_2(L - z_1 - \lambda_{12}/4)} = 0.$$

There is adiabatic invariant $\Sigma/\omega = \mathrm{const}$ [14, 19-21], where Σ is the total light energy stored in the resonator. The simplest explanation is based on the fact that the same invariant takes place for each photon. Since the number of photons is discrete and, therefore, is unchanged at an adiabatic process, the energy Σ of the light field is proportional to the resonance frequency ω or $k = \omega/c$. In this case we have

$$\frac{d\Sigma}{dL} = \frac{-n_2\Sigma}{n_1 z_1 + n_{12}\lambda_{12}/4 + n_2(L - z_1 - \lambda_{12}/4)}. \tag{5}$$

The density of the energy W of a travelling light wave in an optical medium increases by n times as compared with the density

$$W_0 = \varepsilon_0 E_0^2/2 \tag{6}$$

in vacuum (the energy of the magnetic field is taken into account also) because the speed of propagation of the light energy in the optical medium decreases by n times. Thus, the density of the light energy in a travelling wave propagating in an inhomogeneous medium depends on n and is given by $W(n)=W_0 n$. In this case the energy stored in the resonator of unit area can be expressed as follows

$$\Sigma = W_0(n_1 z_1 + n_{12}\lambda_{12}/4 + n_2(L - z_1 - \lambda_{12}/4)].\qquad(7)$$

The force per unit area applied to the right reflector of unit area at $x=L$ is given by $F=-d\Sigma/dL$. Since the standing wave in the resonator consists of two travelling waves propagating in opposite directions, the energy of the standing wave is increased by two times. Taking into account these notes and eqs, (5-7), we obtain that the force per unit area applied to the right reflector is given by

$$F_R = -d(2\Sigma)/dL = 2W_0 n_2. \qquad (8)$$

Repeating the same consideration for the left reflector, we obtain $F_L = -2W_0 n_1$. As is seen, $|F_L| \neq |F_R|$. Since a sum of all internal forces applied to resonator is equal to zero, we obtain that the force applied to the sensor is given by

$$F = -2W_0(n_2 - n_1) = -\varepsilon_0 E_0^2(n_2 - n_1) \qquad (9)$$

This force is independent on position z_1 of $\lambda/4$ plate. The force is produced by two travelling waves propagating in opposite directions. The force produced by a single travelling wave, which density of the energy in vacuum is W_0, can be presented as follows:

$$F = -W_0(n_2 - n_1) = -\varepsilon_0 \frac{E_1^2 n_1}{4}(n_2 - n_1) - \varepsilon_0 \frac{E_2^2 n_2}{4}(n_2 - n_1) =$$
$$-\varepsilon_0 \frac{E_1^2}{4}(\varepsilon_{12} - \varepsilon_1) - \varepsilon_0 \frac{E_2^2}{4}(\varepsilon_2 - \varepsilon_{12})$$

$$(10)$$

This relation can be treated as follows. The first term describes the force applied to the boundary between ε_1 and ε_{12}. The second term describes the force applied to the boundary between ε_{12} and ε_2. There are no forces applied to homogeneous fragments of optical medium. Forces are applied to boundaries between mediums with diverse reflective indexes. Thus, the force applied to a boundary between optical mediums with refractive indexes n_L and n_R located in a field of light wave, which amplitude of the strength of electrical field is equal to E, can be expressed as

$$F = -\varepsilon_0 \frac{E^2}{4}(n_R^2 - n_L^2) = -\varepsilon_0 \frac{E^2}{4}(\varepsilon_R - \varepsilon_L) \qquad (11)$$

Let us now show that the same result can be obtained in a field of standing wave where the square of the electrical field E^2 depends on z accordingly eq.(1) as follows $(4E_0^2/n_1)\text{Sin}^2(kn_1z)$. In accordance with eq. (11) the optical force per unit area at point z_1 is equal $F(z_1) = -(4E_0^2/\varepsilon_1^{1/2})\varepsilon_0(\varepsilon_{12}-\varepsilon_1)\text{Sin}^2(kn_1z)/4 = F_1\,\text{Sin}^2(k\,n_1z_1)$, where F_1 is given by

$$F_1 = -\varepsilon_0 E_0^2 (n_2 - n_1) \qquad (12)$$

The optical force at point z_2 is equal to
$F(z_2) = -(4E_0^2/\varepsilon_2^{1/2})\varepsilon_0(\varepsilon_2-\varepsilon_{21})\quad \text{Sin}^2(kn_1z_1+kn_{12}\lambda/4)/4 \quad = \quad F_1$
$\text{Sin}^2(kn_2z_1+\pi/2) = F_1\text{Cos}^2(k\,n_1z_1)$.

A sum of forces at points z_1 and z_2 is equal to F_1, which is independent on z_1 and is equal to the force obtained on the base of the energetic approach in Eq.(9). Analogously, a sum of forces at points z_3 and z_4 is independent on z_3 and is equal to $-F_1$. As was expected, a sum of forces at four points is equal to zero and is independent on neither z_1 nor w. The sensor is expanded by these forces only.

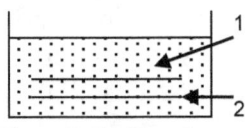

Figure 2. Position of a plane optical resonator in a liquid transparent optical medium

2.3 Optical electrostriction pressure in an optical medium inserted in a plane resonator

Let us previously calculate for the sake of check the electrostriction pressure produced on a liquid or gaseous dielectric in a static electrical field. Similar calculation is presented in many textbooks, in particular, in [1-9]. We present one of possible approaches which is analogous to the approach used below at calculation the electrostriction pressure in a plane resonator. Let a plane capacitor be filled by a liquid dielectric of the permittivity ε, as is shown in figure 2. The energy stored in the capacitor is equal $\Sigma = \varepsilon_0\varepsilon E^2/2 = D^2/(2\varepsilon_0\varepsilon)$, where D is the electrical induction. The charge Q and, therefore, D is preserved in an isolated capacitor with changing ε. In this case, there is an adiabatic invariant $\varepsilon\Sigma$ and, therefore, $d\varepsilon/\varepsilon + d\Sigma/\Sigma = 0$. The pressure on the dielectric is given by

$P=-d\Sigma/dV= - (\Sigma/\varepsilon)d\varepsilon/dV$, where V is the volume of a part of the capacitor of unit area, Σ is the energy stored in the volume. As is seen in Fig.2, the volume in the plane capacitor is unchanged but the density of the dielectric is changed at compression. Since the density τ of the dielectric is in inverse proportion with its volume V, we have $d\tau/\tau=-dV/V$ and the electrostriction pressure can be presented as follows

$$P = - (\Sigma/V)(\tau/\varepsilon)\, d\varepsilon/d\tau =- \tau\, d\varepsilon/d\tau\, (\tau/\varepsilon)W = - \tau\, d\varepsilon/d\tau\, \varepsilon_0 E^2/2$$

$$(13)$$

where $W=\Sigma/V$ is the density of the electric energy. Sign minus corresponds to compression of the medium unlike a conventional positive pressure of gas, which tends to expand the gas.

Let us consider a change in the light energy stored in a plane resonator with a change in the reflective index n of an optical medium that fills the resonator completely as is shown in Fig.2. As is well-known, the eigen frequency of the resonator ω is in inverse proportion with n and product ωn is an adiabatic invariant. On the other hand, the energy Σ stored in a conservative closed oscillating system is proportional to the frequency of oscillations at a slow change in parameters of the system [14, 19-21]. In this case product $n\Sigma$ is an adiabatic invariant also and $dn/n+d\Sigma/\Sigma=0$, where Σ is the light energy stored in the resonator. As a result, a change in n is accompanied by a change in the light energy. The pressure produced by a light wave on the optical medium is given by $P=-d\Sigma/dV=(\Sigma/n)dn/dV$. Since the density τ of the optical medium is in inverse proportion with its volume V, we have $d\tau/\tau=-dV/V$ and the optical electrostriction pressure averaged over space can be presented as follows

$$<P>= -dn/d\tau\, (\tau/n)(\Sigma/V)= -dn^2/d\tau\, (\tau/n^2)(\Sigma/V)/2=$$

$$=-d\varepsilon/d\tau\, (\tau/\varepsilon)(\Sigma/V)/2= -d\varepsilon/d\tau\, (\tau/\varepsilon)\, W/2, \qquad (14)$$

where $W=\Sigma/V$ is the density of the light energy averaged over time and space. Comparing eqs. (13) and (14), one can see that the pressure P in the plane resonator is smaller by two times than that in the capacitor provided their densities of energy are identical.

The density of the light energy in the resonator in accordance with Eq.(1) is the following

$W=4[\varepsilon_0\varepsilon E^2/2 Sin^2(knz)Cos^2(\omega t)+\mu_0\mu H^2/2\, Cos^2(knz)Sin^2(\omega t)]$.

Averaging W over time, we obtain that W is independent on z and is equal to $\varepsilon_0\varepsilon E^2$. Seemingly, the pressure P is equal to the pressure $<P>$ in Eq.(14) and is independent on z also. However, a careful consideration shows that this hasty conclusion is invalid.

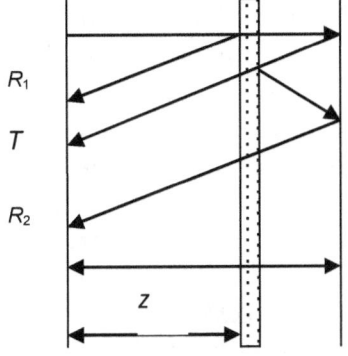

Figure 3. Main (T) and secondary (R_1, R_2) waves in a plane resonator where a tester is embedded

Let a tester in a form of a thin layer of optical medium of Δh thickness and $n+\Delta n$ reflective index ($\Delta h<<\lambda$, $\Delta n<<1$) be imbedded in the plane resonator as is shown in fig. 3. Since the thickness of the tester is small, small waves of amplitudes R_1 and R_2 to first order in Δh arise at reflection from the tester. The index of reflection from the tester is the following [7]

$$R=\frac{r_{LEFT}+r_{RIGHT}\,Exp(-j2kn\Delta h)}{1+r_{LEFT}r_{RIGHT}\,Exp(-j2kn\Delta h)},\text{where } k=\omega/c,\; r_{LEFT}\cong-\Delta n/2n,$$

$r_{RIGHT}=-r_{LEFT}$. Taking into account that $2kn\Delta h<<1$, we obtain $R\cong-jk\Delta n\Delta h$.

Let the resonance takes place in the resonator at the eigen frequency ω and the phase shift at bypassing resonator by light wave $T\,Exp[-i(\omega t-knz)]$ is equal to $-2knL=-2\pi$. An insertion of the tester in the resonator at the distance x from the left reflector violates the resonance at frequency ω. In this case the phase shift becomes different from 2π and can be calculated as follows.

Taking into account fig.3, the following expressions for phases of main T and reflective $R1$, $R2$ waves at $z=0$ can be derived at neglecting phases to first order of smallness

$T=Exp[-j2k\Delta n\Delta h]$

$R_1=|R|Exp[-j(knz-\pi/2+knz+\pi)]=-j|R|Exp(-j2knz)$

$R_2=|R|Exp[-j(knh+\pi+kn(h-z)-\pi/2+kn(h-z)+\pi+knh+\pi]=$
$-j|R|Exp(j2knz)$

The complex amplitude of a sum of oscillations is given by

$$T + R_1 + R_2 = 1 - j2k\Delta n\Delta h - jk\Delta n\Delta h[Exp(-j2kz) + Exp(j2kz)] =$$
$$1 - j2k\Delta n\Delta h[1 - Cos(knz)]$$

Since $2k\Delta n\Delta h \ll 1$, a change in the phase shift at bypassing of the resonator is the following $\Delta\varphi = -2k\ \Delta h\Delta n\ [1-Cos(2knz)] = -4k\Delta h\Delta n Sin^2(knz)$.

Since the resonance is preserved at insertion of the tester, an increase in the phase shift owning an increase in Δn is compensated by a decrease in the resonance frequency ω by $\Delta\omega$. This takes place if $\Delta\omega/\omega = \Delta\varphi/2\pi = -4k\Delta h\Delta n Sin^2(knz)]/2\pi$. Recalling that $\Delta\omega/\omega = \Delta\Sigma/\Sigma$ and $2knL = 2\pi$, we have that a decrease in the energy $\Delta\Sigma$ per init area in plane xy of the resonator is given by $\Delta\Sigma = -2\Sigma\Delta n Sin^2(knz)]/nL = -2W\Delta n Sin^2(knz)]/n$, where $W = \Sigma/L$ is the density of the light energy averaged over z. This change in the energy must be equal to the work Δw performed in compressing the optical medium. The work done per unit volume is given by $\Delta w = -P_{ES}\Delta\tau/\tau$, where τ is the density of the medium, P_{ES} is the electrostriction pressure of the medium due to the presence of electromagnetic oscillations. Since $\Delta\Sigma = \Delta w$ at accepted assumptions, the electrostriction pressure is given by

$$P_{ES} = 2\frac{W}{n}\tau\frac{dn}{d\tau}Sin^2(knz). \quad \text{Since} \quad \frac{dn}{d\tau} = \frac{d\varepsilon}{d\tau}\frac{1}{2n}, \quad \text{we have}$$

$$P_{ES} = W\frac{\tau}{\varepsilon}\frac{d\varepsilon}{d\tau}Sin^2(knz).$$ Averaging P_{ES} over z, we obtain the averaged pressure given by Eq.(14).

The total energy stored in the resonator is independent on time. The electric energy is transformed into magnetic one and vice versa in each period of oscillations. We can calculate the total energy at the time instant when the total energy is concentrated in an electrical form. Averaging the density of the energy $W = \varepsilon_0\varepsilon E_{max}^2 Sin^2(knz)/2$ over z, we have $W = \varepsilon_0\varepsilon E_{max}^2/4$, where E_{max} is the maximal strength of electrical field in the resonator. As a result, we have

$$P_{ES} = \varepsilon_0\tau\frac{d\varepsilon}{d\tau}\frac{E_{max}^2}{4}Sin^2(knz) = \varepsilon_0\tau\frac{d\varepsilon}{d\tau}\frac{E^2(z)}{4} \qquad (15)$$

The same result can be obtained for a ring resonator. In this case the density of the light energy is identical along the ring and is given by $W=\varepsilon_0\varepsilon E^2/2$. Substituting this expression in Eq.(14), we obtain Eq.(15).

3. Optically induced forces calculated on the base of electrostatic approach

Density forces applied to a dielectric in electrostatics at a steady state [1-9] are given by

$$f = \rho E - grad(\varepsilon)\frac{\varepsilon_0 E^2}{2} + grad(\tau\frac{d\varepsilon}{d\tau}\frac{\varepsilon_0 E^2}{2}) - grad(P_H)$$

(16)

The first term describes the Coulomb interaction and is equal to zero at accepted assumptions. The second term describes the *Maxwell density force* in an inhomogeneous dielectric. In a particular case of an abrupt boundary between two dielectrics with permittivities ε_1 and ε_2 the term describes the *Maxwell force per unit of area* given by

$$F = -(\varepsilon_2 - \varepsilon_1)\frac{\varepsilon_0 E^2}{2}$$

(17)

The third term in eq. (16) is the gradient of the *Helmholtz electrostriction pressure* given by

$$P = -\tau\frac{d\varepsilon}{d\tau}\frac{\varepsilon_0 E^2}{2}$$

(18)

The fourth term is the gradient of the *hydrostatic pressure* that compensates the electrostriction pressure. At a steady state a sum of the third and fourth terms is equal to zero.

Comparing eqs. (17) and (11), one can see that the Maxwell force in electrostatics is greater by two times than the force in a field of light wave provided that the amplitude of the alternate strength of electrical field is equal to the strength of a static electrical field. Comparing the electrostriction pressure given by eq. (15) and the electrostriction pressure given by eq.(18) one can see that the same picture. Eqs. (11) and (15) have been derived from energetic considerations. The same result can be obtained by means of averaging E^2 in eq. (16) over time. Similar averaging is used in [14, 22]. Taking into account these notes, formulas of electrostatics can be used for calculation of optical density forces in a field of light waves.

4. Optically induced forces derived from the Maxwell stress tensor

Maxwell stress tensor (MST) in electrostatics in a form of $T_{\alpha\beta}=\varepsilon_0\varepsilon(E_\alpha E_\beta-|E|^2/2)$ can not be used for calculation of optical density forces because $T_{\alpha\beta}$ is derived from Eq. (16) on assumption that rot(E)=0 [4, 5]. This is valid in electrostatics but this is inconsistent in electrodynamics where rot($\boldsymbol{E}$)=-d$\boldsymbol{B}$/dt. MST in electrodynamics in a form of $T_{\alpha\beta}=\varepsilon_0\varepsilon(E_\alpha E_\beta-\delta_{\alpha\beta}|E|^2/2) + \mu_0\mu(H_\alpha H_\beta-\delta_{\alpha\beta}|H|^2)/2$ ought to be used [14]. In this case terms connected with the magnetic field appears. Taking into account Eq.(1), one can see that only diagonal elements are different from zero and

$$T_{zz}=-\varepsilon_0\varepsilon(4E_0^2)\text{Sin}^2(knz)/2-\mu_0\mu(4H_0^2)\text{Cos}^2(knz)/2= -\varepsilon_0\varepsilon E^2,$$

(19)

where $\varepsilon=\varepsilon_1$, $n=n_1$, $E^2=E_0^2/n_1$ at $z<z_1$ and $\varepsilon=\varepsilon_2$, $n=n_2$, $E^2=4E_0^2/n_2$ at $z>z_2$ (fig. 1b). The pressure on the right reflector can be calculated through the surface integral $P = \oint_S \vec{T}dS$, where S represents the closed surface limited by planes S_0, S_1 and planes $x=x_0$, $x=-x_0$, $y=y_0$, $y=-y_0$, where $4x_0y_0=1\text{m}^2$. The only non-vanishing contribution to the integral will come from S_1. Indeed, electrical and magnetic fields at S_0 are absent and contributions from surfaces $x=x_0$, $x=-x_0$, $y=y_0$, $y=-y_0$, are compensated mutually because 1 D structure is considered and fields along x- and y- axes are independent on x and y. Thus, taking into account Eq.(19), we have $F=-2\varepsilon_0(n_2^2)(E_0^2/n_2)=-2\varepsilon_0n_2E_0^2$. The force is directed along the outward normal. The sign is changed if the force is directed along the z axis. Averaging E_0^2 over time and taking into

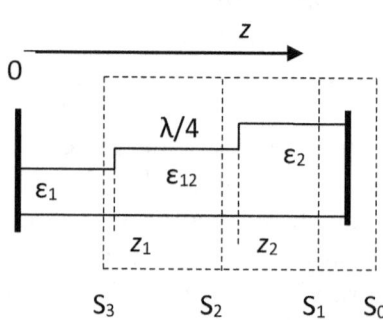

Figure 4. Positions of closed surfaces for calculation of optical forces by MST method

account Eq.(6), we obtain $F=2n_2W_0$. This corresponds to Eq.(8) derived from energetic considerations. As is seen from Eq.(19), F is independent on z. This means that there are no density forces applied to the homogeneous optical medium in interval $z_2<z<L$ in spite of inhomogeneous electrical field.

Integrating MST in Eq.(19) along the closed surface limited by planes S_0 and S_3 (figure 4) we obtain force $F=2n_1W_0$ applied to the right reflector and $\lambda/4$ plate. This means that the force applied to $\lambda/4$ plate is negative and equal to $F=-2(n_2-n_1)W_0$. The same result can be obtained by integrating MST in Eq.(19) along the closed surface limited by planes S_1 and S_3. It is worthwhile to note that MST elements are proportional to the corresponding components of the momentum of light.

5. Optical momentums in optical mediums with different reflective indexes

Having known a force applied to $\lambda/4$ plate, we can calculate a change in the momentum of light penetrating in the optical medium with the reflective index n_2 through $\lambda/4$ plate. As is well-known, the momentum of a travelling light wave with the energy Σ_0 in vacuum is given by $p_0=\Sigma_0/c$. The momentum of light in vacuum of unit area transmitted through $\lambda/4$ plate per one second is given by

$$p_0=(W_0c)/c=W_0, \tag{20}$$

where in accordance with eq. (6) $W_0==\varepsilon_0E_0^2/2$ is the density of the light energy (including magnetic energy) in free space averaged over time. The summary force applied to right reflector due to penetrating a light wave through $\lambda/4$ plate from vacuum accordingly to eq.(8) is $F=2W_0n_2$. This force is equal to a change in the momentum per one second, that is $F=2\Delta p_m=2p_m$, where p_m is the momentum of a travelling light wave propagating in the optical medium. In this case we have $p_m= W_0n_2 = n_2p_0$. As is seen, the Minkowski momentum takes place in the optical medium at penetration of light through $\lambda/4$ plate. The momentum of a wave propagating in the optical medium with the reflective index n_1 is equal to n_1p_0. Then the force applied to $\lambda/4$ plate through which two travelling wave are propagating in opposite directions is equal

$$F = -2(n_2p_0 - n_1p_0) = -2W_0\,(n_2 - n_1). \tag{21}$$

Comparing Eqs. (21) and (9), one can see that the pressure calculated by means of analysis of a change in the momentum (CM) approach is equal to the pressure calculated on the base of the energetic approach. The momentum of a wave is changed when the wave penetrates through the boundaries of different optical mediums only. The momentum is not changed in a homogeneous optical medium. Therefore, no optical density forces arise in a homogeneous medium.

Let us calculate the force applied to an inhomogeneous optical medium which reflective index increases gradually from n_1 to n_2 by means of analysis of a change in the momentum. If the momentum in vacuum of light propagating through unit area per one second is equal to $p_0 = \varepsilon_0 E_0^2/2$, the momentum in an optical medium with the reflective index n is np_0 and a change in the momentum at propagation of light without reflections from the medium with the reflective index n_1 to the medium with the reflective index n_2 is given by $(n_2 - n_1)p_0$. The net force F applied to the inhomogeneous medium of 1 m^2 area is equal to

$$F = -(n_2 - n_1)p_0 = -(n_2 - n_1)\varepsilon_0 E_0^2/2. \tag{22}$$

As is seen, F in eq.(22) is equal to P in eq. (9). Thus, the approach based on an analysis of a change in the momentum is in agreement with our conclusion that the optical density forces are absent in a homogeneous optical medium and are connected with a rise of reflections.

6. Optically induced forces calculated on the base of the Lorentz law

The *Lorentz density force* used in works of Gordon and Loudon [10-12] is given by

$$f = (P \cdot \nabla)E + \dot{P} \times B \tag{23}$$

where $P = (\varepsilon - 1)E$ is the polarization of the dielectric.

The *Lorentz density force* used in works of Mansuripur [13] is presented in the following form at assumption that $\mu = 1$

$$f = -(\nabla \cdot P)E + \dot{P} \times B \tag{24}$$

The first term in eq.(23) is equal to zero in the plane resonator because in accordance with eq. (1) the y- and z- components of the electrical field E are absent and $dE_x/dy=0$. The first term in eq. (24) describes the force which is parallel to the strength E. The z-component of the force is equal to zero in a plane resonator. The second terms of the Lorentz force presented in eqs.(23, 24) are identical. The z-component of $[dP/dt \times B]$ is given by $f_z = 4\varepsilon_0(\varepsilon-1)kE_0^2 Sin(kn_1 z)Cos(kn_1 z)Sin^2(\omega t)$. Time-averaging f_z gives

$$f_z = k\varepsilon_0(\varepsilon-1)E_0^2 Sin(2kn_1 z) \qquad (25)$$

As is seen, f_z is proportional to $Sin(2nkz)$. Integrating $Sin(2nkz)$ over z along width w of the sensor, we obtain that the net force applied to the sensor of unit area is proportional to $Sin^2(kn_1 z_1 + \pi/2 + kn_2 w) - Sin^2(kn_1 z_1 + \pi/2)$. Thus, a net force applied to the sensor on the side of the Lorentz density forces is different from zero. This contradicts to our results obtained above on the base of the independent energetic approach. The energetic approach gives us an assurance that the net force is calculated correctly independently on types of optically induced forces applied to the optical medium, in particular, independently on a specificity of internal forces within the sensor. As is known, a net force of internal forces is equal to zero. Thus, any of two forms of the Lorentz force [10-13] does not give the zeros net force. Both forms are inconsistent.

7. Optical forces applied to an inhomogeneous optical medium which reflective index is changed gradually along propagation of a travelling light wave

Let us next analyze a change in the net force applied to the optical medium where the refractive index is changed along the z-axis slowly and monotonously so that $\lambda dn/dx \ll 1$ and $dn/dz > 0$. As is known, a travelling wave propagates in such medium practically without reflections because reflected waves at points z and $z+\lambda/4$ are compensated mutually. In the same time, E^2 in the travelling wave is in inverse proportion with the reflective index provided all other parameters are unchanged. Indeed, the density of the electrical energy W of a light wave in an optical medium increases by n times as compared with the density $W_0 = \varepsilon_0 E_0^2/2$ in vacuum because the speed of propagation of the light energy in the optical medium decreases by n times. From relation $W = nW_0$ we have $E^2 = E_0^2/n$. The Maxwell force applied to the inhomogeneous part of the optical medium in interval (z_1, z_2) in accordance with eq.(16) is given by

$$F = -\varepsilon_0 \int_{z_1}^{z_2} \frac{d\varepsilon}{dz} \frac{E^2(z)}{4} \, dz = -\varepsilon_0 \int_{\varepsilon_1}^{\varepsilon_2} \frac{E_0^2}{4\sqrt{\varepsilon}} \, d\varepsilon = $$
$$-\varepsilon_0 \frac{E_0^2}{2} (\sqrt{\varepsilon_2} - \sqrt{\varepsilon_1}) = -\varepsilon_0 \frac{E_0^2}{2} (n_2 - n_1) \qquad (26)$$

Reflections arise in an inhomogeneous medium only. Secondary waves reflected from inhomogeneities of optical medium with the slowly changing reflective index are compensated mutually. On the contrary, changes in the momentum connected with the reflected waves are not compensated. The changes are accumulated and the total momentum gradually increases with increasing the reflective index from n_1 to n_2.

8. Discussion

The energetic approach reveals only two types of optically induced force applied to an optical medium in a plane resonator. These are the *Maxwell-like forces* and the *Helmholtz-like electrostriction pressure*. These types are known for a long time. It is shown that classical formulas of electrostatics for the *Maxwell forces* and for the *Helmholtz electrostriction pressure* in a dielectric can be used for calculation of forces and pressure in an optical medium embedded in a field of light waves. As one would expect, the square of the strength of electrical field E^2 in the classical formula of electrostatics [1-9] should be substituted by a half of the square of the amplitude of the strength of electrical field in electrodynamics.

Thus, one can say that a nature of optically induced forces is identical to that in electrostatics although this conclusion has been derived from perfectly different grounds. In this case we can learn in more detail about properties of density forces in electrostatics. An inhomogeneity of the strength of an electrical field in a plane resonator enables to clarify a specificity of striction forces connected with the *Helmholtz electrostriction pressure* in a homogeneous optical medium and an inhomogeneous electrical field. It has been shown on the base of an energetic approach that there are no density forces in this case at a steady state. This is an unexpected result is supported by results of CM and MST approaches. It turns out that this result contradicts notions existing above century as well as investigating intensively at present.

Actually, there is a sum of four terms in eq.(16). The first term is absent at accepted assumptions. The fourth term is the gradient of the hydrostatic pressure that is independent on the light wave and can not be considered as an optically induced force. The hydrostatic pressure is a subsequent reaction of the optical medium on a rise of the optically induced forces. The second term describes density forces in an inhomogeneous medium. The net force of these density forces can be different from zero. For example, the net forces applied to $\lambda/4$ plates in fig. 1a, 1b are different from zero. The net force described this term in a homogeneous medium is equal to zero.

The net force of the third term can be different from zero in an inhomogeneous electrical field and homogeneous optical medium. Really, as is seen from Eqs.(1, 16), a contribution of the third term of Eq.(16) in a total net force applied to the homogeneous region of w width in Fig.1a is different from zero and is equal to

$$F = -\tau \frac{d\varepsilon}{d\tau} \frac{\varepsilon_0 E^2}{4} [Sin^2(kn_{12}w + \varphi) - Sin^2\varphi]$$

, where φ is the phase at $z=z_2$. This contradicts to our result that a net force applied to the region of w width is equal to zero and our notions that a net force of any pressure is equal to zero. This contradicts also to our results obtained from CM and MST approaches.

The contradiction can be resolved as follows. Internal and external density forces ought to be not mixed in the same expression. A net force of internal density forces is equal to zero because they arise owning a tendency of the medium to be compressed (expanded). In this case each internal density force is

63

accompanied by its counterparts in accordance with the third Newton law. But such counterparts are absent for internal density forces presented by Eq.(16).

A separate analysis of internal and external density forces enables to exclude this drawback. A change to analysis of pressures instead of internal density forces enables to consider both internal density force and its counterpart simultaneously. Indeed, there are both the electrostriction and hydrostatic pressures at any point of the medium. Signs of these pressures are different. Their sum should be equal to zero in any point at a steady-state. The gradient of the sum is equal to zero also. The third and fourth terms should be excluded from Eq.(16) and Eq.(16) becomes

$$f_1 = \rho E - grad(\varepsilon)\frac{\varepsilon_0 E^2}{2} \qquad (27)$$

where f_1 is the external density force applied to the dielectric located in a static electrical field of E strength. The electrostriction pressure should be taken into account separately in accordance with Eq.(15). Certainly, Eqs (19) and (15) should be considered simultaneously as source equations for compressible optical mediums. In this case a change in the permittivity ε due to compression of the optical medium by the electrostriction pressure should be taken into account.

As is seen, all considered approaches for calculation of optically induced force give identical results. The forces obtained on the base of different approaches are not summed. Seemingly, perfectly different notions and laws show themselves in these cases. Nevertheless, we are forced to recognize that various approaches give the same magnitude of optically induced forces in spite of various physical notions that explain a rise of these forces.

Exception is for approaches based on the Lorentz forces presented in [10-13]. These approaches give incorrect result that density forces in a homogeneous optical medium embedded in an inhomogeneous field of light wave are different from zero.

9. Conclusion

There are only two types of optically induced force applied to an optical medium in a plane resonator. These are the *Maxwell-like forces* and the *Helmholtz-like electrostriction pressure*. These types are known for a long time. It is shown that classical formulas of electrostatics for the *Maxwell forces* and for the *Helmholtz electrostriction pressure* in a dielectric can be used for calculation of forces and pressure in an optical medium embedded in a field of light waves. As one would expect, the square of the strength of electrical field E^2 in the classical formula of electrostatics should be substituted by a half of the square of the amplitude of the strength of electrical field in electrodynamics.

It is shown also that there are no optically induced forces applied to a homogeneous optical medium located in an inhomogeneous electrical field of light wave. This conclusion is valid also for a homogeneous dielectric located in an inhomogeneous static electrical field. Contrary to generally accepted notions, the gradient of the *Helmholtz electrostriction pressure* in an inhomogeneous electrical field does not entail a rise in the optically induced forces at a steady-state. The inhomogeneous *Helmholtz electrostriction pressure* is compensated in each elementary volume by the corresponding inhomogeneous hydrostatic pressure. This pressure ought to be taken into account along with the *Helmholtz electrostriction pressure*. As a result, a sum of these pressures at a steady state is equal to zero. Thus, the *Helmholtz electrostriction pressure* is responsible for compression of the medium but is not responsible for a rise of the density forces at a steady state.

Optically induced forces applied to reflectors of a plane resonator can be calculated on the base of energetic, MST and CML approaches. Since these forces arise at reflection of light wave, they can be called by *radiation forces*. The forces of the same type arise at reflection from $\lambda/4$ plate. In this case they can be considered as the *Maxwell-like forces*. Thus, the *Maxwell-like forces* can be considered in optics as a particular case of the *radiation forces*.

Analysis of a change in the momentum of light in a homogeneous optical medium shows that there is no change in the momentum at a steady-state and, therefore, there are no *Maxwell-like forces* applied to the medium. There is the inhomogeneous *Helmholtz-like electrostriction pressure* and hydrostatic pressure only. The same conclusion was derived from the energetic and MST approaches. In this connection an incorporation of the gradient of the *Helmholtz electrostriction pressure* in the formula for *Maxwell forces* leads to a tangle. Two separate equations for external and internal optically induced forces ought to be used.

The approaches based on the Lorentz forces give incorrect results because accordingly them a net force is different from zero in a homogeneous optical medium. This contradicts to results obtained on the base of energetic, ES, MST and CML approaches.

References

[1] Physical encyclopedia, Ed. A M Prokhorov vol.3 pp 32-33 (Moscow, Bol'shaia Rossiyskaia encyclopedia, 1992) (In Russian).

[2] L D Landau, E.M., Lifshits and L P Pitaevskii, *Electrodynamics of Continuous Media* (Oxford, Heinemann, 1984).

[3] M. Abraham and R. Becker *Theorie der Elektrizitat* (Teurner, Leipzig,1932) Band 1 p.111.

[4] J. A. Stratton *Electromagnetic theory* (Mc.Graw-Hill, N.Y., 1941) p.146.

[5] W. K. H. Panofsky and M. Phillips *Classical electricity and magnetism* (Addison-Wesley, Mass., 1960).

[6] I. E. Tamm, *Fundamentals of theory of electricity* (Nauka, Moscow.1966) (In Russian).

[7] D. V. Sivukhin 2004 *Electricity* (Moscow, Fizmatlit) (in Russian)

[8] Landsberg G S 1976 *Optics* (Nauka, Moscow)

[9] J D Jackson 1999 *Classical electrodynamics*, (Wiley, N.Y.)

[10] Gordon J P 1973 *Phys. Rev. Lett.* **30** 139

[11] Loudon R 2002 *J Mod. Opt.* **49** 812

[12] Loudon, R Barnett S.M. 2006 *Optics Express* **14** No. 24 11855

[13] Mansiripur M 2004 *Optics Express* **13** 5375

[14] Rakich P T, Popovic M A, Wang Z Z 2009 *Optics Express* **17** No. 20 18116

[15] Rakich P T, Davids P, Zweng Wang 2010 *Optics Express* **18** No 14 14439

[16] Novotny L and Hecht B *Principles of Nano-optics* (Cambridge University Press, 2006)

[17] Pfeifer R N C, Heckenberg T A and Rubinsztein-Dunlop H 2007 *Rev.Mod.Phys.* **79** 1197

[18] Barnett S M and Loudon R 2010 *Phil. Trans. R. Soc.* **A 368** 927

[19] Torchigin V P 1996 *J. of Tech. Phys.* (USSR) **66**(4) 128

[20] Povinelli M L, Loncar M, Ibanescu M, Smythe E, Capasso F, and Joannopoulos J, 2005 *Optics Express* **13** 8286

[21] Torchigin V P and Torchigin A V 2005 *Eur. Phys. J. D.* **32** 385

[22] Boyd R W 2003 *Nonlinear Optics* (Academic Press)

Propagation of a light pulse inside matter in a context of the Abraham-Minkowski dilemma

V. P. Torchigin, A.V. Torchigin

Institute of Informatics Problems, Russian Academy of Sciences, Nakhimovsky prospect, 36/1, Moscow, 119278, Russia, tel +7 499 1332532, email: v_torchigin@mail.ru

PACS: 42.65.Jx, 42.65.Tg

Keywords: optical forces; momentum density; momentum flux density; Minkowski momentum; Abraham momentum, Abraham force.

Abstract

It is shown that a light pulse propagating in an optical medium exerts the optical pressure on the medium in the regions where leading and trailing edges are propagating. This effect is derived from analysis of unambiguous thought experiments which results contradict one other. It is shown that a magnitude of the pressure is equal to $(n-1/n)W_0$, where n and W_0 is the refractive index of the medium and the momentum flux density of the same pulse in free space, respectively. The Abraham form of the momentum of light is redundant if the optical pressure is taken into account. In this case the dilemma disappears because one of the rival alternatives disappears.

1. Introduction

There have been extensive debates about the correct expression of the momentum density of electromagnetic waves in linear media for more than 100 years since the original papers of Minkowski [1] and Abraham [2], and even so there is still some confusion or at least disagreement among authors. In 2010 various arguments in favor of Abraham and Minkowski theories were presented in review [3] devoted to the 100-year anniversary of this problem. This is not the place to review the large literature devoted to this problem. The interested reader can study the recent review by Pfiefer *et al.* [4].

There is an opinion that the the Balazs thought experiment [5] uniquely select the Abraham momentum as a momentum of the field in matter. A behavior of a transparent block of an optical medium through which a light pulse is propagating is considered. The main argument in favor of the Abraham momentum is that the behavior of the block is explained correctly on assumption that the Abraham momentum takes place within the block. Minkowski momentum would predict a motion of the block in the opposite direction to the incident pulse. Most recently, Barnett and Loudon reanalyzed the controversy and argued that both momenta are "correct" because both can be measured, but in different situations [3]. Following the analysis of Balazs and repeating arguments of the thought experiment, they concluded that "it is difficult to see how any component of our derivation could seriously be open to question". "If argument advanced in favor of the Abraham momentum were to be incorrect, than that would bring into question uniform motion of an isolated body as expressed in the Newton's first law of motion".

Other thought experiments can be used to clear up a situation about kinds of optically induced forces (OIF) and their interaction. One of them is considered in [6] on the basis of the energetic approach where no previous assumption about a number and kinds of OIF is done. It is shown that the momentum flux density of a continuous light wave in an optical medium increases by n times as compared with that of the same wave in vacuum. This corresponds to the Minkowski approach.

An assumption is advanced that the difference between magnitudes of the momentum is connected with a difference between properties of a light pulse with leading and trailing edges and continuous light wave. For example, the wavelength of a light pulse is decreasing gradually because the electrostriction pressure is produced by a light pulse and a part of its energy is transmitted to the medium in a form of elastic energy [7]. Besides, it is shown that additional density forces applied the regions where leading and trailing edges of the pulse are propagating should arise. This conclusion is derived from an attempt to coordinate rival results of thought experiments which are based only on laws of conservation of the momentum and energy. The additional forces are equal to the known Abraham density force which existence is debated for a long time.

2. Momentums of light in an optical medium based on thought experiments

Pros and Cons of the Minkowski and Abraham forms of the momentum of light are analyzed in [3]. A single decisive argument in favor of the Abraham form is the Balazs thought experiment [5]. It is believed that the most direct way to calculate the momentum of a photon in a medium is to use the Newtonian idea that the centre of mass (or more precisely the centre of mass-energy) of an isolated system undergoes uniform motion. For this purpose a situation presented in Fig.1 is considered. Here light pulse 1 propagates in vacuum at speed of light c and its momentum is equal to p_0. When the light pulse enters the transparent medium in a form of a block 2, its speed slows to c/n and, as a result, it takes the time $T=nL/c$ to travel through the block, where L and n are the thickness and refractive index of the block, respectively. It is supposed that the block begins to move to the right when the pulse enters the block and stops its motion when the pulse exits the block. Actually, a sum of the momentums of the light pulse and block is preserved and, as a result, the momentum of the light pulse within the block decreases. When the light pulse exits the block, the momentum of the light pulse is recovered and the momentum of the block becomes equal to zero. As a result, the block stops. It is shown that the light pulse should transmit to the block momentum $p_0(1-1/n)$ [3].

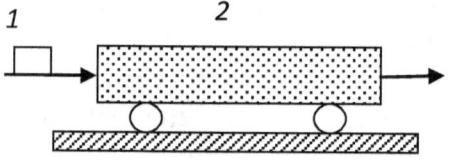

Fig. 1 Propagation of light pulse 1 through block 2 of optical medium with the reflective index n

Denoting the momentum of light inside the block by x, we have from the law of conservation of the momentum $x+ p_0(1-1/n)=p_0$ or $x=p_0/n$. As was noted, only the conservation of momentum and the uniform motion of the centre mass-energy in deriving this result is used and it is difficult to see how any component of this derivation could seriously be open to question. If the momentum flux density of the same light pulse propagating in free space is equal to W_0, the pressure (force per unit area) produced by the light pulse on the block is given by

$$P_A = W_0(1 - 1/n)_.$$

(1)

The same result we obtain on assumption that the momentum flux density of the light pulse within the block is equal to W_0/n

On the other hand, there is a thought experiment based on the energetic approach where no assumption is done about kinds of optically induced force applied to the block [6]. In accordance with the thought experiment the pressure produced on the block by a continuous light wave is given by

$$P_M = -W_0(n - 1)$$

(2)

The same result we have on assumption that the momentum flux density of the wave within the block is equal to $W_0 n$

3. Elimination of contradictions between thought experiments

Thus, the momentum of the light pulse in the block corresponds to the Abraham form and is smaller by n times than that in free space. The momentum of a continuous light wave in the block corresponds to the Minkowski form is greater by n times than that in free space. A reason of a discrepancy is required to disclose. The reason should be connected with properties of a light pulse that are different from that of a continuous light wave. Unlike a continuous light wave, there are leading and trailing edges in a light pulse. Let us assume that additional pressure should arise at these edges at a propagation of a light pulse in a homogeneous optical medium. Two pressures should be applied to the regions of the block where the leading and trailing edges are located at given time instants. Besides, these pressures should provide zero net force applied to the optical medium if the pulse is propagating in a homogeneous optical medium. A net pressure is different from zero when the light pulse enters or exits the block. In this case only one pressure is applied to the block. The pressure is applied in a time interval equaled to the duration of the light pulse. Thus, these two

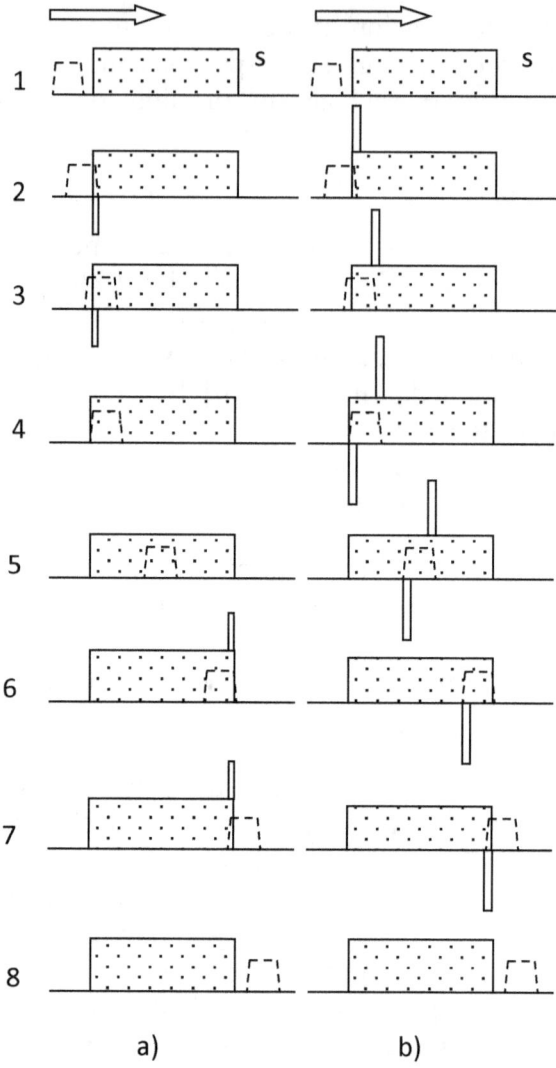

additional pressures P_{add} on the edges of the light pulse can produce the pressure that overcome the negative pressure given by Eq. (2) and provide a positive displacement of the block. An action of the additional pressures P_{add} is shown in Fig.2b along with action of the main pressures P_M given by Eq. (2) and is shown in Fig.2a.

A joint action of the main and additional pressures should provide a replacement of the block in the direction required by the thought experiment. In this case the total pressure applied to the block in

Fig. 2. Serial stages of propagation of a light pulse (dashed trapeze) through optical block (dotted rectangular). a) pressures P_M arising on the front and faces of the block; b) pressures P_{add} arising at leading and trailing edges of the light pulse.

accordance with Eq. (1) should be equal to $W_0(1-1/n)$. Then from the relation $-(n-1)W_0 + P_{add} = W_0(1-1/n)$ we have

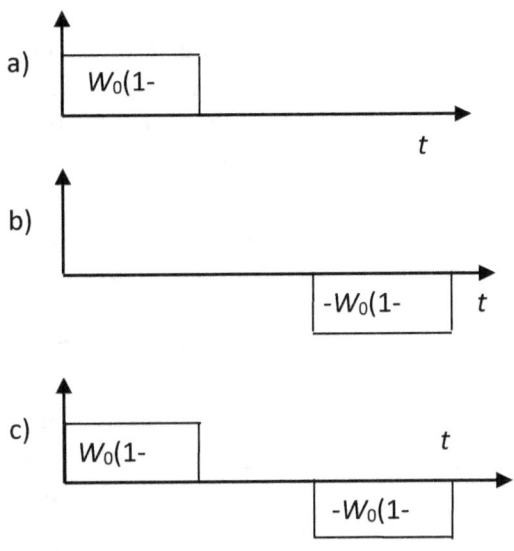

$$P_{add} = W_0(n - 1/n)$$

(3)

Fig. 3. Dependence on time of pressures produced by light pulse on
a) front face,
b) back face,
c) total pressure on the block
on assumption that the momentum of light pulse corresponds to the Abraham form and additional pressure is absent.

A travel of the light pulse along the block is shown in Fig.2b. As is seen, there are positive and negative pressures P_{add}. In this case a replacement of the block is positive as it is expected in accordance with the thought experiments.

We have two alternatives for explanations of the thought experiments. The first one is a traditional explanation: the momentum of the light pulse decreases in an optical medium by n times. This conclusion is generalized for any light waves and a notion about the Abraham momentum is introduced. In this case dependence on time of pressures exerted by the light pulse on front and back faces and total pressure exerted on the block is shown in Fig.3.

In accordance with the second alternative the momentum of a light pulse corresponds to the Minkowski form that is valid also for continuous light waves (when amplitudes of light waves are independent on time). When a light pulse is propagating, additional pressures arise in an optical medium where edges of the light pulse are located. When the light pulse is entering the block, the momentum of the light pulse is increasing by n times in accordance with the Minkowski form. The pressure $P_M = -(n-1)W_0$ applied to the front face of the medium arises. In the same time the additional pressure $W_0(n-1/n)$ applied to the region of the block

73

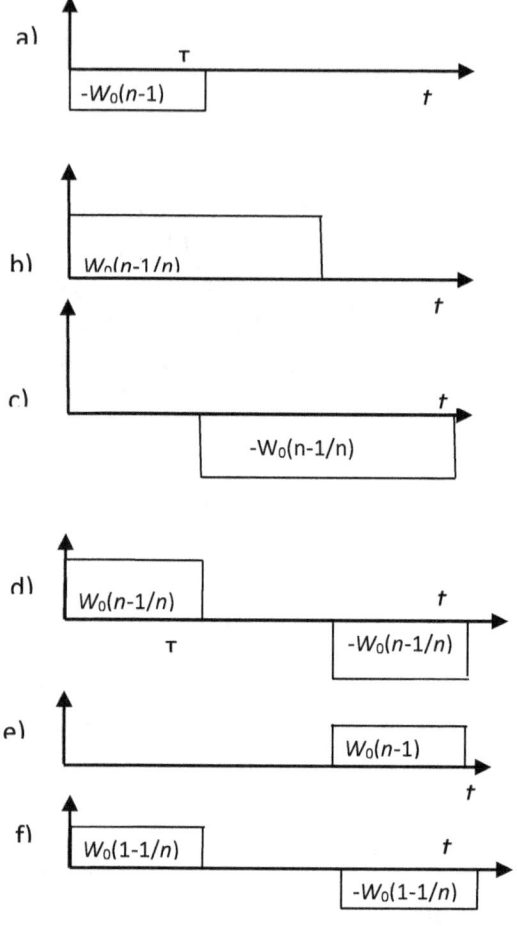

Fig. 4. Dependence on time of pressures
a) on front face,
b) produced by leading edge along the block,
c) produced by trailing edge along the block,
d) a sum of additional pressures,
e) on back face,
f) total pressure on the block
on assumption that the momentum of light pulse
corresponds to the Minkowski form and the
additional pressures arise at propagation of the light
pulse along the block. Duration of the pulse is τ.

where the leading edge of light pulse is located arises. As a result, a total pressure $W_0(-n+1+n-1/n=W_0(1-1/n)$ applied to the optical medium arises (Fig.4). An impression could be created that the momentum flux density of light is changed from W_0 to W_0/n. When the light pulse is exiting the medium, pressure $W_0(n-1)$ due to a change of the momentum on the back face of the block arises. In the same time additional pressure - $W_0(n-1/n)$ produced by the trailing edge of pulse is not compensated by the pressure on the leading edge of the pulse. As a result, there is pressure $-W_0(1-1/n)$ applied to the block. The pressure stops the block. Thus, a behavior of the block can be explained both on an assumption that there is the Abraham form of the momentum and on an assumption that there is the Minkowski form of the momentum along with the additional pressures produced by edges of a light pulse.

4. Thought experiment on reflection of a light step from an ideal reflector

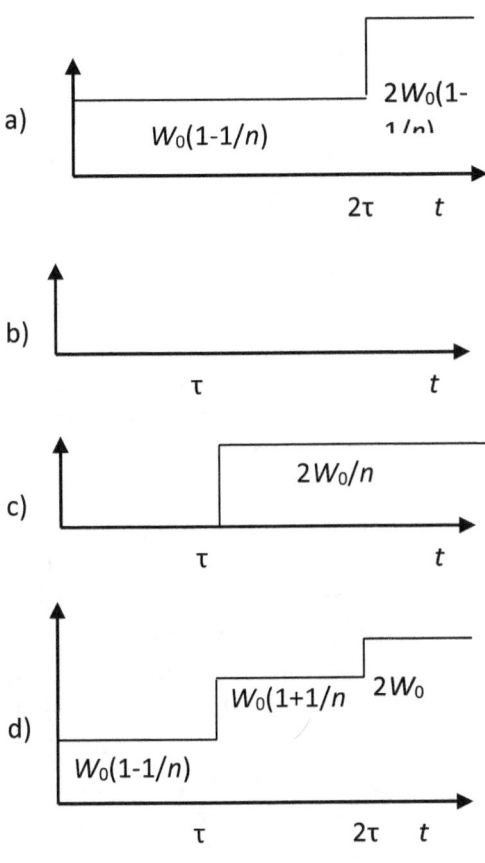

a)

$W_0(1-1/n)$

$2W_0(1-1/n)$

2τ t

b)

τ t

c)

$2W_0/n$

τ t

d)

$W_0(1-1/n)$

$W_0(1+1/n)$ $2W_0$

τ 2τ t

Fig. 5. Dependence on time of pressures produced by a step of light on
a) front face,
b) along the block (additional pressure),
c) on reflector,
d) total pressure on the block
on assumption that the momentum of light corresponds to the Abraham form and the additional pressure is absent.

To choose the most suitable alternative, between the "Abraham alternative" in Fig.3 and the "Minkowski alternative" in Fig.4 let us consider one more thought experiment where a light step of the energy density W_0 propagates in vacuum and enters the block at $t=0$ where it is propagating in an optical medium along the z axis and is reflecting from an ideal reflector located on the back face of the block. Unlike a light pulse, there is only one edge in the step. If the momentum flux density of the light step corresponds to the Abraham form and is equal to (W_0/n) in the Abraham alternative, the pressures on the front face, back face and total pressure on the block are shown in Fig. 5. The step of light wave enters the block at $t=0$, reflects from the reflector at $t=\tau$ and exits the block at $t=2\tau$.

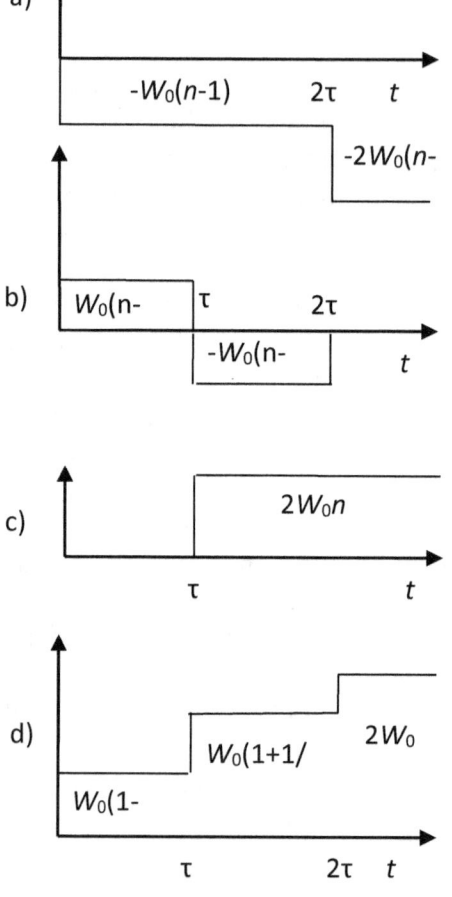

a) $-W_0(n-1)$ 2τ t

$-2W_0(n-$

b) $W_0(n-$ τ 2τ

$-W_0(n-$ t

c) $2W_0n$

τ t

d) $W_0(1+1/$ $2W_0$

$W_0(1-$

τ 2τ t

Fig. 6. Dependence on time of forces exerted by a step of light on
a) front face,
b) along the block (Abraham force),
c)reflector,
d) total force applied to the block on assumption that the momentum of light corresponds to the Minkowski form and the Abraham forces arise at propagation of the edge of the step along the block.

In accordance with the Minkowski alternative the momentum of a light step corresponds to the Minkowski form that is valid for continuous light waves also (when amplitudes of light waves are independent on time). In this case the additional pressures arise in an optical medium where the edge of the light step is propagating (Fig.6). When the light step is entering the block, the momentum of the light step is increasing by n times in accordance with the Minkowski form. The pressure $P_M = -(n-1)W_0$ on the front face of the medium arises. In the same time the additional pressure $W_0(n-1/n)$ applied to the medium where the leading edge of light pulse is located arises. As a result, a total pressure $W_0(-n+1+n-1/n=W_0(1-1/n)$ on the optical medium arises. This looks like as if the momentum flux density of light is changed from W_0 to W_0/n.

As is seen from Figs. 5, 6, the pressures on the block at $t > 2\tau$ are equal to $2W_0$ independently on the forms of the momentum. It follows from the conservation of total momentum. The momentum flux density in vacuum is changed by

76

$-2W_0$ and the total pressure on the block is equal to $2W_0$. The pressures on the block at $0<t<\tau$ in Fig. 5 and 6 are identical.

However, at $t>2\tau$ there is a essential difference between the Abraham and Minkowski alternatives. Pressures on the reflector and on the front face of the block for the Minkowski alternative are different from that for the Abraham alternative. In accordance with Fig.5 these pressures are both positive. In accordance with Fig. 6 the pressure on the front face of the block is negative. The pressure on the reflector is greater by n times than the pressure on the reflector by the same wave propagating in vacuum.

There is an experimental confirmation of this fact. When a mirror was immersed in a dielectric medium, the radiation pressure exerted on the mirror was proportional to the reflective index of the medium. An accuracy of this effect was 1.2% [8]. These results correspond to that obtained on the basis of the energetic approach where no assumption about kinds of OIF and forms of the momentum of light is done [6].

The same conclusion can be derived from the fact that at $t>2\tau$ there is a steady state process within the block where any changes of amplitudes of light waves and phase relations between them are absent. In this case in accordance with the second thought experiment [6] the pressure on the front face is negative and is equal to $-(n-1)W_0$, the pressure on the reflector is positive and is equal to $2nW_0$. The same pressures are shown in Fig. 6.

Thus, the inclusion of the additional pressures along with the Minkowski form of the momentum of light into consideration of propagation of light pulses enables one to coordinate results obtained, on one hand, from the Balazs thought experiment and, on the other hand, from the numerous evidence in favor of the Minkowski momentum. No notion about the Abraham momentum of light is required in this case. An existence of the additional forces on edges of a light pulse has been derived from unambiguous thought experiments.

5. A nature of the additional pressure.

Let us discuss a possible physical origin of these additional pressures. It is reasonable to assume that so called Abraham density force can produce the additional pressures. A magnitude of these forces is given by

$$F_A = \frac{d}{dt}\{\frac{n^2-1}{c^2}[E \times H]\} = \frac{d}{dt}\{\frac{n^2-1}{c}\varepsilon_0 E^2\} = \frac{d}{dt}\{\frac{n^2-1}{c}2W\}$$

(4)

where E, H and W are the electrical and magnetic field strength and the energy density of the light pulse, respectively [9-12]. On assumption that $W=0$ at $t<0$ and $W=W_0 Sin^2(\omega t)$ at $t>0$ we obtain from Eq. (4) that F_A is oscillating at very great frequency 2ω and F_A averaged over period of oscillation is equal to zero. However this is a misconception.

Let $W(t)$ be an arbitrary continuous differentiable function of time t. In this case F_A can be presented in accordance with Eq.(4) as follows

$$F_A = \frac{n^2-1}{c}\{2W(t)\omega Sin2(\omega t) + 2\frac{dW(t)}{dt}Sin^2(\omega t)\} \cdot \qquad \text{Averaging}$$

$W(t)$ over period of oscillation, we have $F_A(t) = \frac{n^2-1}{c}\frac{dW_0(t)}{dt}$.

Density force F_A is the force per unit volume [N/m³]. The pressure P_A [N/m²] is produced by the density forces applied to the leading edge of the light pulse in region $0<z<\Delta l$, where Δl is the length of the edge, $W(0)=0$, $W(\Delta l)=W_0$. In this case

$$P_A = \int_0^{\Delta l} F_A dz = \frac{n^2-1}{c}\int_0^{\Delta l} F_A \frac{dz}{dt}dt =$$

$$\frac{n^2-1}{c}\frac{c}{n}\int_0^{\Delta W} dW = (n-1/n)W_0$$

(5)

Comparing Eq. (3) and Eq.(5), one can see that the Abraham force F_A produces on the block the same pressure as our expected pressure P_{add} given by (3). Thus, the Abraham forces in the thought experiments along with forces that arise due to a change of the momentums provide a transmission to the block momentum (W_0/n) rather than $(W_0 n)$. An existence of the additional pressures has been derived from thought experiments. Analogously, an existence of the Abraham density force can be derived from an existence of the additional pressures or from the thought experiments.

Thus, on assumption that the momentum of a light pulse in an optical medium increases by n times as compared with that in vacuum, like the momentum of a continuous light wave does, a conclusion can be derived on the basis of an analysis of pressures in the thought experiments that there should be additional pressures applied to an optical medium when a light pulse is propagating in the optical medium. The pressures are applied to the regions where leading and trailing edges of the light pulse are propagating. The density forces that produce the pressures are equal to the Abraham density forces that arise in an optical medium at a propagation of light pulses. Thus, an existence of the Abraham density force is derived from results of thought experiments. A notion about the Abraham momentum is redundant in this approach. Only one form of the momentum exists. There are no rival alternatives to form the dilemma.

6. Conclusion

A propagation of a light pulse in an optical medium is accompanied by arising of optically induced density forces in the region where leading and trailing edges of the pulse are propagating. The force applied to the region where the leading edge is propagating is directed along a direction of the propagating. The density forces are equal to the Abraham forces. These forces produce pressures which magnitude is equal to $(n-1/n)W_0$. For comparison, the pressure produced by the same light pulse propagating in free space is equal to $2W_0$. A sum of pressures produced at the leading and trailing edges of a light pulse is equal to zero. This effect enables one to give another explanation of the Balazs thought experiment. No notion about the Abraham momentum is required. There is only the Minkowski form of the momentum in an optical medium and the Abraham density forces. A motivation for introducing of a notion about the Abraham momentum disappears completely. Since the rival Abraham form of the momentum is absent, the dilemma disappears also.

References

[1] H. Minkowski, Die Grundgleichungen fur die elektromagnetischen Vorgange in bewegten Korpern, Nachr. Konigl. Ges. Wiss. Gottingen, (1908) 53-111.

[2] M Abraham, Zur Elektrodynamik bewegter Korper, Rend. Circ. Mat. Palermo **28**, (1909) 1-28.

[3] S. M. Barnett and R. Loudon, The enigma of optical momentum in a medium, Phil. Trans. R. Soc. A (2010) 368-927.

[4] R. N. C. Pfeifer, T A Nieminen, N.R. Heckenberg and H. Rubinsztein-Dunlop. Momentum of an electromagnetic wave in dielectric media, Rev.Mod.Phys. **79(4)** (2007) 1197-1216 .

[5] N.L. Balazs, The energy-momentum tensor of the electromagnetic field inside matter, Phys. Rev. 91 (1953) 408-411.

[6] V. P. Torchigin and A. V. Torchigin, Comparison of various approaches to the calculation of optically induced forces, Annals of Physics **327** (2012) 2288-2300.

[7] V. P. Torchigin and A. V. Torchigin, An increase in the wavelength of light pulses propagating through a fiber, Phys. Lett. A 311 (2003) 21-25.

[8] R. V. Jones and J. C. S. Richard, The pressure of radiation in a reflecting medium, Proc. R. Soc. London Ser. A 221 (1954) 480-498.

[9] J. D. Jackson *Classical electrodynamics*, (Wiley, N.Y., 1999).

[10] L. D. Landau, E.M., Lifshits and L P Pitaevskii, *Electrodynamics of Continuous Media* (Oxford, Heinemann, 1984).

[11] J. A. Stratton *Electromagnetic theory* (Mc.Graw-Hill, N.Y., 1941).

[12] W. K. H. Panofsky and M. Phillips *Classical electricity and magnetism* (Addison-Wesley, Mass., 1960).

Interrelation between various types of optically induced forces

V. P. Torchigin, A.V. Torchigin

Institute of Informatics Problems, Russian Academy of Sciences, Nakhimovsky prospect, 36/1, Moscow, 119278, Russia, tel +7 499 1332532, email: v_torchigin@mail.ru

PACS: 42.65.Jx, 42.65.Tg

Keywords: optical electrostriction; optical forces; Maxwell stress tensor; Lorentz density force; Helmholtz pressure.

Abstract

Optically induced forces applied to a transparent optical medium are analyzed. It is shown on the base of various approaches that the density of optically induced forces applied to a homogeneous optical medium located in an inhomogeneous electrical field is equal to zero at a steady-state. This result contradicts to that obtained by means of an approach based on the Lorentz density force. An explanation is presented that the Lorentz density force is compensated at a steady-state by other type of optically induced force. Thus, a calculation of optically induced force based on the approach using the Lorentz force is inconsistent

1. Introduction.

With recent advances in the fabrication of nano-sized devices [1], optical effects that are too small to have a measurable impact on a macroscopic scale are attracting increasing interest. The mass and the dimensions of optical devices have been miniaturized to the degree that device tuning through optical actuation is possible at micro- to milli-watt power levels [2-6]. The technological possibilities of such system can be explored in light driven mechanically variable systems that can perform trapping [7, 8], actuation [4, 6], transduction [2, 3, 5] and manipulation [7] of nanoscale objects. Since the mechanical state of such systems is intimately linked to its optical state, these mechanical functions can lead to variable directional couplers [4, 9], parametric optical processes in cavites [2-5, 10], ultra-widely tunable microcavities [7].

In the same time generally accepted terms for designating optical forces (optically induced forces) are absent. In parallel with terms *"gradient force"*, *"electrostrictive force"*, *"radiation pressure"* [11] we can meet terms *"electrostrictive pressure"*, *"Lorentz density force"*, *"optical density force"*, *"optically induced force"*. Besides, there are various approaches for calculation of optically induced forces.

One of them supposes that results obtained in electrostatics [12-18] can be used for calculation of optically induced forces (ES approach). This assumption is based on the fact that forces in electrostatics are proportional to the square of the strength of electrical field E^2. In this case the square of the strength of alternate electrical field of light wave averaged over cycle is different from zero and can be used instead of the E^2. There are no mentions about the magnetic field of light waves.

Another approach is based on the *Maxwell stress tensor* (MST). It is supposed that the knowledge of the full electromagnetic field distribution in mechanically variable optical systems is a prerequisite for the computation of optical forces. In these electromagnetic field-based calculations, MST is numerically integrated over a closed surface surrounding the moveable components in the system to compute the optical forces acting on them [11]. In this case the magnetic field of light waves is taken into account. A term *"radiation pressure"* is used for designation of this type of forces. Recently in 2010, MST approach was supplemented by the *electrostrictive components of optical forces* [19]. A term *"electrostrictive forces"* was introduced along with term *"radiation pressure"*. It is supposed that just as with the MST, *the electrostrictive force* can be found from *the electrostriction pressure* distribution.

Seemingly, an approach based on an analysis of a change in the momentum of light in an optical medium (CM approach) could be used as an etalon. However, at present there is no general accepted notion about a magnitude of the momentum. Mincowsky and Abraham proposed two different approaches to this problem. A great deal of controversy has centered on the rival claims of the

Abraham expression for the photon momentum hv/n where n is the refractive index of the medium. The Минлowski momentum is greater by n^2 times. The few available experiments seem to favor the latter, while many theories favor the former. This is not the place to review the large literature devoted to this problem. The interested reader can study the recent review by Pfeiffer et al. [20]. In 2010 various arguments in favor of Abraham and Minkowski theories were presented in review [21] devoted to the 100-year anniversary of this problem.

In 1974 Gordon initiated the idea of calculating the *Lorentz density forces* (force per unit volume acting on the body) [22]. Later this approach was studied by many investigators, in particular, Loudon, Barnett and Mansuripur [23-25]. It is believed "that the Lorentz force provides the fundamental description of radiation pressure effects. It has the advantage that no prior assumptions are made about the magnitude of the optical momentum in the medium. Moreover, the calculated force represents the measured quantity, and not some subsidiary quantity that may not itself be directly measurable". Besides, an advantage of this approach is "the simplicity and safety of calculations based on the Lorentz force and the dangers of calculations based on derived expressions involving elements of the Maxwell stress tensor, whose contributions may vanish in some situations but not in others" [23]. For example, in a series of recent publications, Mansuripur has shown that "the pressure and momentum of the electromagnetic field can be obtained directly from the Lorentz law of force in conjunction with the macroscopic equations, without the need to determine the stress tensor" [26]. Moreover, he has proposed a new form of the stress tensor that "differs from the well-known tensors of Abraham and Minkowski which have been at the centre of a century-old controversy surrounding the momentum of the electromagnetic field in transparent materials".

Thus, each approach deals with own types of optically induced forces. This is *gradient density forces* and the *electrostrictive pressure* in ES approach. This is *the radiation pressure* in MST approach. This is *optically induced forces* connected with a change in the momentum of light in CM approach. This is the *Lorentz density forces* in approaches used in [23-26]. The question arises.

What types of forces should be taken into consideration in a general case? Are types listed above independent or they are different terms connected with features of a specific case? To answer this question optically induced forces in optical medium embedded in a plane optical resonator have been calculated on the base of energetic approach where no assumptions are done about a physical origin of these forces [27-28]. It turns out that results obtained on the base of this approach are identical to that obtained on the base of all mentioned above approaches with one exception [29].

In spite of various terms and approaches used for calculation of optically induced forces, there are only two types of forces [29]. These are density forces arising in an inhomogeneous optical medium where its permittivity is changed in space and the electrostriction pressure which takes place in any medium. The first type of forces can be designated as gradient forces because they arise in the medium where gradient of the permittivity ε (or the refractive index $n=\varepsilon^{1/2}$) is different from zero. The same type of forces in an optical medium is responsible for the radiation pressure. The pressure arises at a boundary of media with different ε as a result of reflection of light wave from the boundary. If a change of ε is continuous, local reflections produce gradient density forces. The second type of forces can be designated as the electrostriction pressure. Unlike the gradient force, which is vector, the electrostriction pressure in liquids or gases is scalar. There is a generally accepted misleading conception that an inhomogeneity in the electrostriction pressure produces the gradient density force which is proportional to the gradient of the electrostriction pressure. It has been shown that the electrostriction pressure is compensated by the hydrostatic pressure at each point of space and no density forces arise [29].

As for the exception that it is concerned to the Lorentz density forces. It turns out that the Lorentz density forces in a homogeneous optical medium located in a plane optical resonator are different from zero. This contradicts results obtained on the base all other approaches. In the same time there are experimental evidence that the Lorentz forces exist. Actually, like the Maxwell force arising at an interaction between electrical current and the magnetic field, there are forces arising at an interaction between the alternate polarization current (alternate polarization) and magnetic field (the Maxwell-like type of force) [30-32].

To resolve this contradiction, we consider the simplest situation where optically induced forces are equal to zero. Minimal mathematics is required in this case. Indeed, to determine whether forces are equal to zero or not is much simpler than to calculate a magnitude of these forces as it was done in [29]. It is shown that there is an additional new type of optically induced forces arising at an interaction between the alternate magnetization and the electric field. Quite recently experimental evidence confirming an existence of this type of forces was published [33].

A methodological character of the paper induces us not to aspire to brevity and to present formulas and expressions well known in the classical macroscopic electrodynamics.

2. Distribution of electrical and magnetic field in a plane optical resonator

The simplest situation is considered where an optical medium is imbedded in a one-dimensional oscillating structure in a form of plane optical resonator consisting of two parallel plane ideal reflectors, between which a plane wave is reflected in serial. The medium is as simple as possible. It is a liquid. It is linear, lossless, motionless, non-dispersion, non-magnetic, homogeneous, isotropic, incompressible, without free charges and currents. Incompressibility is required to neglect the elastic energy stored in the optical medium. A steady state is considered. In this case relations between amplitudes and phases in all points of the medium are independent on time and optically induced density forces are independent on time also. This situation is typical for works devoted to optically induced forces.

On assumption that the z axis is perpendicular to the reflectors and plane $z=0$ coincides with a plane of one reflector, a field of a plane standing wave in a plane resonator filled by a homogeneous optical medium with the permittivity $\varepsilon>1$ and permeability $\mu>1$ is given by

$$E_x = 2E_0\eta^{-1/2}Sin(knz)Sin(\omega t), E_y = 0, \quad E_z = 0,$$
$$H_x = 0, H_y = 2E_0\eta^{1/2}Z_0^{-1}Cos(knz)Cos(\omega t), \quad H_z = 0$$

$$(1)$$

where E_0 is the amplitudes of the strength of electrical field of a travelling wave propagating in vacuum, $n=(\mu\varepsilon)^{1/2}$ is the refractive index, $\eta=(\varepsilon/\mu)^{1/2}$, $Z_0=(\mu_0/\varepsilon_0)^{1/2}$. The electrical and magnetic fields in their maximums are increased by two times as compared with that in the corresponding travelling wave. The length of the resonator is equal to $L=(\lambda/2)N$, where $\lambda=2\pi/(kn)$ is the wavelength of the light wave, N is an integer, $k=\omega/c$, ω is the angular frequency of the light wave.

3. Optically induced forces calculated on the base of the Lorentz density force

At present there are two alternative formalisms advanced by Gordon [22] and Mansuripur [25]. The Lorentz density force used in works of Gordon and Loudon [22, 23] is given by

$$\boldsymbol{f} = (\boldsymbol{P}\cdot\nabla)\boldsymbol{E} + \dot{\boldsymbol{P}}\times\boldsymbol{B}$$

$$(2)$$

appropriate to a dipolar medium. Here $\boldsymbol{E}$ and $\boldsymbol{B}$ are respectively the electric field strength and the magnetic induction associated with the light radiation, $\boldsymbol{P}=(\varepsilon-1)\boldsymbol{E}$ is the polarization of the optical medium.

The Lorentz density force used in works of Mansuripur [26] at assumption that $\mu=1$ is presented in the following alternative form

$$\boldsymbol{f} = -(\nabla\cdot\boldsymbol{P})\boldsymbol{E} + \dot{\boldsymbol{P}}\times\boldsymbol{B}$$

$$(3)$$

appropriate to a medium formed from individual charges. At present both forms are studied simultaneously.

As is seen, the first term in Eq. (2) is equal to zero in the plane resonator because in accordance with Eq. (1) the y- and z-components of the electrical field E are absent and $dE_x/dx=0$. The first term in Eq. (3) describes the force which is parallel to the strength E. The z-component of the force is equal to zero in a plane resonator.

The second terms $\dot{\boldsymbol{P}} \times \boldsymbol{B}$ in Eqs. (2, 3) are identical. We have from (1, 2) that

$\dot{P}_x = 2\omega\varepsilon_0(\varepsilon-1)\eta^{-1/2}E_0 Sin(knz)Cos\omega t$ and

$B_y = 2E_0\eta^{1/2}(\mu_0\varepsilon_0)^{1/2}Cos(knz)Cos\omega t$.

In accordance with (2, 3) we have

$f_z = \dot{P}_x B_y =$
$4\varepsilon_0(\varepsilon-1)(\omega/c)E_0^2 Sin(knz)Cos(knz)Cos^2(\omega t)$.

Time-averaging f_z gives the following expression for the density force in a plane resonator

$$f_z = (\omega/c)\varepsilon_0(\varepsilon-1)E_0^2 Sin(2knz). \qquad (4)$$

As is seen, the density force is depend on z and is different from zero in a plane resonator.

4. Optically induced forces derived from various approaches

Let us show on the base well-known approach connected with an analysis of a change in the momentum of light that the density of optically induced forces in a homogeneous optical medium inserted in a plane resonator is equal to zero. It is well known that the momentum of a travelling light wave propagating in a homogeneous optical medium is unchanged. This conclusion is valid for the travelling waves propagating in any directions as well as for the standing wave produced by two travelling waves propagating in opposite directions. It is surprising for us why nonzero optically induced forces in a homogeneous optical medium do not cause bewilderment. For example, one can see graphs, where a change of the density of optically induced force along a homogeneous optical medium is presented [34]. A change of the density of optically induced force along ●/4 plate (anti-reflection coating layer) is presented in [26]. Undoubtedly, the optical medium within ●/4 plate is homogeneous.

Let us next show that the same result takes place on the base of an energetic approach. Let a sensor in a form of a plate of w width with the permittivity $\varepsilon > 1$ and permeability $\mu = \varepsilon$ is embedded in the empty resonator. The wave impedance of the sensor is equal to that of the vacuum because $\mu/\varepsilon = 1$. In this case there are no reflections from the sensor. Let there be a resonance in the resonator with the inserted sensor and, therefore, the phase shift of light wave at bypassing the resonator is equal $2\pi N$, where N is an integer. Since there are no reflections from the sensor, the phase shift is not changed with moving the sensor along the z axis. As a result, conditions of the resonance are preserved and the eigen frequency ω of the resonator is not changed.

As is known there is adiabatic invariant $\Sigma/\omega=$const [27-28] in a closed lossless oscillating system, where Σ is the total energy stored in the system, ω is its eigen frequency. The simplest explanation is based on the fact that the same invariant takes place for each photon. Since the number of photons is discrete and is unchanged at an adiabatic process, the energy Σ of the light field is proportional to the resonance frequency ω. Since the energy stored within the resonator is unchanged with changing sensor's position, the mechanical work produced at moving the sensor is equal to zero. Then the net force applied to the sensor is equal to zero at any its position and width w. This means that the density of optically induced force applied to the sensor is equal to zero.

Let us now show that the same conclusion can be derived from an analysis of the Maxwell stress tensor which is given by $T_{\alpha\beta}=\varepsilon_0\varepsilon(E_\alpha E_\beta-\delta_{\alpha\beta}|E|^2/2) + \mu_0\mu(H_\alpha H_\beta- \delta_{\alpha\beta}|H|^2/2)$, where $\delta_{\alpha\beta}$ is Kronecker's delta [11, 12]. Taking into account Eq.(1), one can see that a single diagonal component $T_{zz} =-\varepsilon_0\varepsilon E_x^2/2-\mu_0\mu H_y^2/2$ is different from zero. Averaging E_x^2 and H_y^2 over time in Eq.(1), we obtain

$$T_{zz}=-[\varepsilon_0\varepsilon(E_x^2)\text{Sin}^2(knz)/4+\mu_0\mu(H_y^2)\text{Cos}^2(knz)/4]= -\varepsilon_0(\mu\varepsilon)^{1/2}E_0^2. \quad (5)$$

The force applied to the volume confined by planes $z=z_0$ and $z=z_{00}$ as well as planes $x=x_0$, $x=-x_0$, $y=y_0$, $y=-y_0$, where $4x_0y_0=1\text{m}^2$, can be calculated through the surface integral [12]

$$F = \oint_S \sum_\beta T_{\alpha\beta} n_\beta dS + \varepsilon_0\mu_0\varepsilon\mu \frac{d}{dt}\int_V S dV$$

, where S represents the closed surface formed by the pointed planes, n_β is the outward normal to the surface, V is the volume confined by surface S, S is the Poynting vector. At a steady-state the second term drops out and the optically induced force can be expressed entirely in terms of the stress tensor at the boundary.

The only non-vanishing contribution to the integral will come from planes $z=z_0$ and $z=z_{00}$. Indeed, contributions from surfaces $x=x_0$, $x=-x_0$ are compensated mutually because 1 D structure is considered in which fields along the z- axis is independent on x and y. The same is valid for surfaces $y=y_0$, $y=-y_0$. Thus, taking into account Eq. (5), we have that the net force applied to the left plane of the volume at $z=z_0$ is given by $F_1=-\varepsilon_0(\mu\varepsilon)^{1/2}E_0^2$. As is seen, F_1 is independent on z_0. This means that the force F_2 applied to the surface $z=z_{00}>z_0$ is equal to F_1 by value. But a direction of the outward normal for the surface $z=z_{00}$ is opposite to that for the surface $z=z_0$. This means that the net force applied to the medium at $z_0<z<z_{00}$ is equal to zero. Since z_0 and z_{00} are arbitrary, there is no density forces applied to any fragment confined by arbitrary planes.

The same result can be obtained from the following classical expression presented in many handbooks and encyclopedias [12]

$$f = \rho E + j \times H - \varepsilon_0 grad(\varepsilon)E^2 / 2 - \mu_0 grad(\mu)H^2 / 2$$

$$(6)$$

Since $\rho=0$, $j=0$, $grad(\varepsilon)=0$, $grad(\mu)=0$ within the resonator, the density force f is equal to zero.

The electrostriction pressure is not taken into account in Eq.(6). The electrostriction pressure is proportional to E^2 and, therefore, depends on z. There is a widespread judgment that an inhomogeneity in the electrostriction pressure entails an appearance of the density force. Indeed, the following formula for density force in a static electrical field can be seen in many textbooks [13-19].

$$f = \rho E - \varepsilon_0 grad(\varepsilon)E^2 / 2 + grad(\varepsilon_0 \tau \frac{d\varepsilon}{d\tau} E^2 / 2) - grad(P_H)$$

$$(7)$$

Here $\varepsilon_0 \tau \dfrac{d\varepsilon}{d\tau} E^2 / 2$ is the electrostriction pressure and its gradient makes a contribution in the density force, τ is the density of the optical medium, P_H is the hydrostatic pressure. As is seen,

four terms in Eq.(7) are equal in rights and describe four density forces. For example, the second and third terms is combined for dilute gas where the density τ is proportional to $(\varepsilon\text{-}1)$ [13]. Their sum is equal to $\varepsilon_0(\varepsilon-1)grad(E^2/2)$. Eq. (7) was derived a lot of years ago. It can be seen in textbook [14] published in 1932

What actually happens is that the electrostriction pressure is produced by inner forces which net force is equal to zero in any volume. For example, the net force applied to the medium in the resonator between planes $z=z_0$ and $z=z_{00}>z_0$ with regard to pressures at planes $z=z_0$ and $z=z_{00}$ is given by

$$\int_{z0}^{z00} \frac{d}{dz}[\varepsilon_0\tau\frac{d\varepsilon}{d\tau}E^2(z)/2]dz + P(z_0) - P(z_{00}) =$$

$$P(z_{00}) - P(z_0) + P(z_0) - P(z_{00}) = 0$$

Since z_0 and z_{00} are arbitrary, a net force applied to any volume is equal to zero. Thus, an inhomogeneity of the electrostriction pressure can not produce the density forces.

5. Discussion

Although Eqs. (2, 3) are incorrect, nevertheless there are experimental evidence [31-33] which confirm an existence of the Ampere-like density force arising at an interaction between polarization currents and an external constant magnetic field. There is no doubt that polarization currents can interact not only with constant magnetic field but also with an alternate magnetic field. Such alternate magnetic field takes place in the resonator. In this case the Ampere-like density force is given by

$$f_A = \dot{P} \times B.$$

$$(5)$$

In accordance with Eq.(5) this density force is different from zero. However, we have shown that there are no optically induced density forces within the resonator. If we recognized an existence of the Ampere-like density forces in a homogeneous optical medium, we need to recognize an existence of other type of forces which counteract the Ampere-like force at a steady-state.

Since the following relations take place at a steady

$$\frac{d}{dt}(\boldsymbol{P} \times \boldsymbol{B}) = \varepsilon_0(\varepsilon - 1)\frac{d}{dt}(\boldsymbol{E} \times \boldsymbol{B}) =$$
$$\frac{(\varepsilon - 1)}{c^2}\frac{d}{dt}(\boldsymbol{E} \times \boldsymbol{H}) = \frac{(\varepsilon - 1)}{c^2}\frac{d\boldsymbol{S}}{dt} = 0 \quad, \tag{6}$$

one can assume that the following density force compensates f_A.

$$\boldsymbol{f}_{AA} = -\dot{\boldsymbol{B}} \times \boldsymbol{P} \quad. \tag{7}$$

Quite recently experimental evidence confirming an existence of this type of forces was published [34]. Unlike the Ampere-like forces in (5), a sum of

$$\boldsymbol{f}_A + \boldsymbol{f}_{AA} = \frac{(\varepsilon - 1)}{c^2}\frac{d\boldsymbol{S}}{dt} \tag{8}$$

averaged over time at a steady state is equal to zero.

Eq.(8) resembles the following formula for the Abraham density force

$$\boldsymbol{f} = \frac{(\varepsilon\mu - 1)}{c^2}\frac{d\boldsymbol{S}}{dt} \quad. \tag{12}$$

Since μ-1<<1 for real optical media, a difference between Eqs. (6) and (12) is negligibly small. However, appearance of index ($\varepsilon\mu$-1) instead of index (ε-1) enables us to expand a notion of 'polarization currents'.

Thus, an approach based on the Lorentz density force can not be considered as an alternative to other approaches.

6. Conclusion

There is no density of optically induced forces at a steady state in a homogeneous optical medium and an inhomogeneous field of light waves in a plane optical resonator. However, the approach based on the Lorentz force in Gordon's and Loudon's form [22, 23] gives a result which contradicts to that obtained on the base various other approaches. The approach based on the Lorentz force in Mansuripur's form [25] gives incorrect results also. Calculation of optically induced forces in a field of light waves on the base of the Lorentz force is inconsistent.

Along with the Ampere-like force, which is a component of the Lorentz force, ought to take into account an additional new type of forces arising at an interaction between the alternate magnetization and the electrical field of a light wave. There is experimental evidence in existence of the new type of forces.

References

[1] L. Novotny and B.Hecht, *Principles of Nano-optics* (Cambridge University Press, 2006).

[2]. T. Carmon, H. Rokhsari, L. Yang, T. Kippenberg, and K. Vahala, Phys. Rev. Lett. **94**, (2005) 223902.

[3]. T. J. Kippenberg, H. Rokhsari, T. Carmon, A. Scherer, and K. J. Vahala, Phys. Rev. Lett. **95**, (2005) 033901.

[4]. M. L. Povinelli, M. Loncar, M. Ibanescu, E. J. Smythe, S. G. Johnson, F. Capasso, and J. D. Joannopoulos, Opt. Lett. **30**, (2005) 3042.

[5] M. Eichenfield, R. Camacho, J. Chan, K. J. Vahala, and O. Painter, Nature **459**, (2009) 550.

[6] M. Eichenfield, C. P. Michael, R. Perahia, and O. Painter, Nature Photon. **1**, (2007) 416.

[7] P. T. Rakich, M. A. Popovi.c, M. Soljaˇci.c and E. P. Ippen, Nature Photon. **1**, (2007) 658.

[8] A. Mizrahi, and L. Schachter, Opt. Express **13**, (2005) 9804.

[9] A. Mizrahi and L. Schchter, Opt. Lett. **32**, (2007) 692.

[10] M. Notomi, H. Taniyama, S. Mitsugi, and E. Kuramochi, Phys. Rev. Lett. **97**, (2006) 023903.

[11] P. T. Rakich, M. A. Popovic, and Zheng Wang Opt. Express **17** no.20 (2009) 18116.

[12] Physical encyclopedia, Ed. A M Prokhorov vol.3 pp 32-33 (Moscow, Bol'shaia Rossiyskaia encyclopedia, 1992) (In Russian).

[13] L D Landau, E.M., Lifshits and L P Pitaevskii, *Electrodynamics of Continuous Media* (Oxford, Heinemann, 1984).

[14] M. Abraham and R. Becker *Theorie der Elektrizitat* (Teurner, Leipzig,1932) Band 1 p.111.

[15] J. A. Stratton *Electromagnetic theory* (Mc.Graw-Hill, N.Y., 1941) p.146.

[16] W. K. H. Panofsky and M. Phillips *Classical electricity and magnetism* (Addison-Wesley, Mass., 1960).

[17] I. E. Tamm, *Fundamentals of theory of electricity* (Nauka, Moscow.1966) (In Russian).

[18] D. V. Sivukhin 2004 *Electricity* (Moscow, Fizmatlit) (in Russian)

[19] P. T. Rakich, M. A. P. Davids and Zheng Wang Opt. Express 18 no.14 (2009) 14439.

[20] R N C Pfeifer, T A Heckenberg and H. Rubinsztein-Dunlop *Rev.Mod.Phys.* **79** (2007) 1197

[21] S M Barnett and R Loudon *Phil. Trans. R. Soc.* **A 368** (2010) 927

[22] J P Gordon *Phys. Rev. Lett.* **30** (1973) 139

[23] R Loudon *J Mod. Opt.* **49** (2002) 812

[24] R Loudon, S.M. Barnett *Optics Express* **14** No. 24 (2006) 11855

[25] M Mansiripur *Optics Express* **13** (2004) 5375

[26] M Mansiripur *Optics Express* **16** (2008) 5193

[27] M L Povinelli, M Loncar, M Ibanescu, E Smythe, F Capasso, and J Joannopoulos, 2005 *Optics Express* **13** 8286

[28] V P Torchigin and A V Torchigin 2005 *Eur. Phys. J. D.* **32** 385

[29] V. P. Torchigin and A. V. Torchigin, Annals of Physics **327** (2012) 2288

[30] G.B. Walker, G Walker Nature" **265** (1977) 324.

[31] G.B. Walker, G Walker, Can. J. Phys. **55** (1977) 2121.

[32] R.P. James, Proc. Natl. Acad. Sci. USA **61** (1968) 1149.

[33] G.L.J.A. Rikken and B.A. van Tiggelen, Phys. Rev. Lett. **108** (2012) 230402.

[34] A.R. Zakharian, M. Mansuripur, J.V. Moloney *Optics Express* **13** No. 7 (2005) 2321.

Resolution of the age-old dilemma about a magnitude of the momentum of light in matter

V. P. Torchigin, A.V. Torchigin

Institute of Informatics Problems, Russian Academy of Sciences, Nakhimovsky prospect, 36/1, Moscow, 119278, Russia, tel +7 499 1332532, email: v_torchigin@mail.ru

PACS: 42.65.Jx, 42.65.Tg

Keywords: optical forces; momentum density; momentum flux density; Minkowski momentum; Abraham momentum, Abraham force.

Abstract

There are two competitive opinions about a magnitude of the momentum of light in matter as compared with that of the same light in free space. The Balazs thought experiment gives unambiguous theoretical evidence based only on laws of mechanics that the momentum of light in matter in smaller by n times than that of the same light in free space where n is the refractive index of the matter. On the other hand, there is other theoretical evidence confirmed by experimental investigation that the momentum of light in matter is greater by n times as compared with that in free space. Thus, a question arises what is a magnitude of the momentum of light in reality. This ancient dilemma can not be resolved till now. We show that the momentum density flux of light in matter is greater by n times than that in free space. In the same time a propagation of a rectangular light pulse in matter considered in the Balazs thought experiment is accompanied by two optical pressures arising in the regions of the matter where leading and trailing edges of the pulse are propagating and the intensity of light is changed in time. This fact gives us an opportunity of alternative interpretation of the Balazs thought experiment that enables one to mach contradictory results of mentioned experiments. Since the contradiction overcomes, the dilemma disappears.

1. Introduction

The momentum of light in a conventional transparent optical medium is one of known unsolved classic surprises in theoretical physics. There have been extensive debates about the correct expression of the momentum density of electromagnetic radiation in linear media for more than 100 years since the original papers of Minkowski [1] and Abraham [2], and even so there is still some confusion or at least disagreement among authors. In accordance with Minkowski, momentum of light in an optical medium increases by n times as compared with the momentum of the same light in free space where n is the refractive index of the medium. In accordance with Abraham, momentum of light in an optical medium decreases by n times. Various arguments in favor of the Abraham and Minkowski forms are presented in reviews [3-6].

There are several approaches in attempts to resolve the dilemma. The oldest and most widespread one is based on the energy-momentum tensor formalism [1, 2, 6]. Beginning in the late 1960's something approaching a consensus emerged [6]: Both the Minkowski momentum and the Abraham momentum are "correct", "but they speak to different issues, and it is largely a matter of taste which of the two (or perhaps even one of the other candidates that have from time to time been proposed) one identifies as the "true" electromagnetic momentum. Only the total stress-energy tensor carries unambiguous physical significance, and how one apportions it between an "electromagnetic" part and a "matter" part depends on context and convenience. Minkowski did it one way, Abraham another; they simply regard different portions of the total as "electromagnetic". Except in vacuum, "electromagnetic momentum" by itself is an intrinsically ambiguous notion. For example, when light passes through matter it exerts forces on the charges, setting them in motion, and delivering momentum to the medium. Since this is associated with the wave, it is not unreasonable to include some or all of it in the electromagnetic momentum, even though it is purely mechanical in nature. But figuring out exactly how and where this momentum is located can be very tricky. One would like to write down, once and for all, the complete and correct total stress-energy tensor—electromagnetic plus mechanical".

This position prevents an analysis of processes connected with transformation of the electromagnetic momentum into mechanical one and vice versa at propagation of light in an inhomogeneous optical medium and prevents to use a powerful method of a calculation a magnitude of optically induced forces (OIF) by means of an analysis of a change of the momentums. Having known a change of the momentum density fluxes in various points of the matter, the density of OIF can be calculated without knowledge a distribution of electrical and magnetic fields.

Last time several papers have been published [7, 8, 10] devoted to this problem. It is declared that the problem has been resolved at last. The Barnett resolution [7] recognizes that "the both forms are correct" and "the Abraham and Minkowski momenta are, respectively, the kinetic and canonical optical momenta". However, this approach can not solve the problem that is formulated by Barnett as follows: "why is it that the experiments supporting one or other of these momenta give the results that they do?" Thus, this resolution gives no possibility to use the approach based on an analysis of a change of the momentum of light in practical applications.

Mansuripur believes also that he has solved the controversy [8]. His solution is based on the generalized expression of the Lorentz force. However this expression is incorrect because it gives incorrect result in simplest cases [9].

Various attempts to resolve the Abraham-Minkowski dilemma is reviewed in the resent paper by Kemp [10] where references to the Barnett and Mansuripur solutions are presented. The following conclusion is derived: "a complete picture of electrodynamics has still yet to be full interpreted".

There are the following real experiments performed by Jones *et al* [11, 12]. When a mirror is immersed in a dielectric liquid, the radiation pressure exerted on the mirror is proportional to the refractive index of the liquid. An accuracy of this effect was 0.05%. On assumption that the light has the momentum that changes its sign at reflection from the reflector, one can conclude that the momentum of the light in the liquid is greater by n times than the momentum of the same light in free space.

However this conclusion contradicts to results obtained in the same time from the Balazs thought experiment [13] based on the generally accepted law of the momentum conservation. It is shown theoretically that a transparent block through which a light pulse is propagating without reflection should be displaced in a direction of the propagation of the light pulse. As a result, the block is moving along the light pulse when the pulse is propagating inside the block. In this case a part of the momentum of the pulse is transferred to the block. Therefore, the momentum of light inside the block is smaller than that of the same pulse propagating in free space. Thus, the momentum of light inside the block corresponds to the Abraham form. As is pointed in review [3], "If argument advanced in favor of the Abraham momentum were to be incorrect, than that would bring into question uniform motion of an isolated body as expressed in the Newton's first law of motion".

Since it is impossible in a frame of the energy-momentum tensor formalism to resolve the controversy, we will consider a controversy between unambiguous results of real and thought experiments and terms connected with names of Abraham and Minkowski we will not use. In this case momentum that is smaller by n times than that in free space we will denote as B-momentum meaning that a magnitude of the momentum is derived from the Balazs thought experiment. Accordingly, the momentum which magnitude is greater by n times in matter than that in free space we will denote by J-momentum meaning that the magnitude is derived from the Jones *et al* experiments [11, 12]. By the way, in accordance with our thought experiment [14] a conclusion was derived theoretically that the momentum of a continuous light wave propagating in matter corresponds to the J-momentum. No assumption about kinds and physical origin of OIF responsible for an increase of the momentum is made. Thus, there are two unambiguous rival thought experiments [13, 14] where no assumption about kinds of optically induced force and their physical origin is made. These experiments are a safe ground for analysis a magnitude of momentums and OIF in matter.

Hitherto all attempts to match rival results of the experiments were failed because a reason of this discrepancy was not found out. Recognition that both results are correct because they correspond to experiment is insufficient. An explanation is required why the momentums are different. We present our explanation and show that it can be presented many years ago because the explanation is based on classical laws of mechanics.

2. Notation and basic definitions

Above all things, it is worthwhile to note that our notion about the momentum of light is taken from mechanics where this notion was introduced several centuries ago for a body of mass M moving at speed v as a product of Mv. In accordance with Newton the momentum characterizes the "quantity of motion". A reason of a change of the momentum p is a force f and in accordance with the second Newton law $dp/dt=f$. In a closed system (one that does not exchange any matter with the outside and is not acted on by outside forces) the total momentum is constant. This fact, known as the law of conservation of momentum, is implied by Newton's laws of motion. [15, 16]. A magnitude of the momentum of a light pulse propagating in vacuum is generally accepted and is given by $\mathcal{E}_{pulse}/c$, where $\mathcal{E}_{pulse}$ is the energy of the light pulse.

If a continuous plane light wave is considered, its momentum is equal to infinity. In this case the momentum of the wave propagating through a cross-section of unit area per unit time is considered. This momentum is equal to the momentum density flux (MDF). MDF of a continuous plane light wave propagating in vacuum at speed c is given by $(W_0 c)/c=W_0$ [J/m^3] where $W_0=\varepsilon_0 E_0^2/2$ is the energy density of the light wave, E is the amplitude of the strength of the alternate electrical field of the light wave. We will call this MDF by the electromagnetic MDF. A mechanical pressure P applied to a body transmits to the body the mechanical MDF equal P [N/m^2=J/m^3].

There are optically induced forces (OIF) produced by the light propagating in an optical medium. As a result, the light interacts with matter (an exchange of the momentums between the light and matter takes place). The law of the conservation of the momentums and the third Newton law are valid at this interaction. As a result, each OIF changes the mechanical MDF of matter. In turn, a counterpart of the OIF (COIF) that arises in accordance with the third Newton law changes the electromagnetic MDF of light. Thus, each interaction is accompanied a redistribution between mechanical and electromagnetic MDFs. A sum of these MDFs is not changed. Thus, OIF is responsible for a change of the mechanical momentum and COIF is responsible for a change of the electromagnetic momentum. Usually, relations between electromagnetic and mechanical MDFs before interaction are known. The mechanical momentum of any light wave in free space is equal zero. Having known a distribution of OIF in space and time, a behavior of the mechanical and electromagnetic MDFs in space and time can be calculated.

3. Momentum of light derived from the Balazs thought experiment

A main argument in favor of the B-momentum in matter is the Balazs thought experiment [13] that is described in last time many times [3-6, 8]. A behavior of a transparent block of an optical medium through which a light pulse is propagating is considered. The behavior is explained correctly on assumption that the B-momentum takes place within the block. The J-momentum would predict a motion of the block in the opposite direction to the incident pulse.

The following 1D structure is considered. A light pulse of a plane light wave is propagating along the z axis in free space at speed of light c. The pulse enters a block of thickness D and refractive index n, is propagating within the block and leaves the block preserving its initial energy and momentum. The optical medium of the block is as simple as possible. The medium is linear, dispersionless, lossless, homogeneous, and nonmagnetic. It is supposed

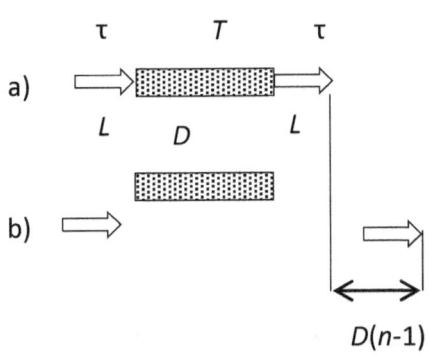

Fig.1. Propagation of the light pulse (a) through the block; (b) outside the block

that measures are undertaken to exclude reflections when the light pulse enters and leaves the block. For example, the block is confined by slabs of thickness d, where $D>>d>>\lambda$ and the refractive index n is changed gradually from 1 to n in the slab before the front face of the block and from n to 1 in the slab after the back face. Here λ is the wavelength of light. As is known, reflections of light wave entering the block through such slab can be neglected.

Fig. 1a shows positions of the light pulse of duration τ and length $L=c\tau$ in a form of hollow arrows before and after penetrating through the block. Fig. 1b shows positions of the same pulse propagating in free space outside the block. A time of propagation of the pulse through the block in Fig. 1a at distance $L + D$ is equal $\tau+T$ where $T= D/(c/n)$ is time of propagation of the leading edge of the pulse through the block. The pulse in Fig. 1b propagates at distance $c(Dn/c+\tau)=Dn+L$ during this time. The distance is greater than the distance $L+D$ in Fig.1a by $D(n-1)$. The condition for an unchanged center of the mass-energy in Fig 1b relatively that in Fig. 1a is the following, $Mc^2\Delta z = W_0 LSD(n-1)$ where M is the mass of the block, Δz is a displacement of the block when the pulse has left the block, W_0 is the energy density of the pulse in free space, S is the area of the cross-section of the block, W_0LS is the energy of the light pulse. In this case

$$\Delta z = W_0 LSD(n-1)/ Mc^2 \tag{1}$$

Since the system consisting of the pulse and block is closed, the momentum conservation law for momenta before and after entrance of the pulse into the block gives

$$W_0 LS / c = Mv_{block} + p_{pulse} \tag{2}$$

where p_{pulse} is the momentum of the pulse inside the block, v is the velocity of the block when the pulse is propagating inside the block. Since the block is shifted by Δz during time interval $\tau+D/(c/n)$, the velocity of the block is $v_{block}=\Delta z/[\tau+D/(c/n)]$. Then from Eqs (1), (2) we have

$$p_{pulse} = W_0\tau[1 - \frac{1-1/n}{1+L/(Dn)}] \tag{3}$$

As is seen, the momentum of the pulse inside the block decreases by n times in a limiting case only when length of the pulse L tends to zero. It is this case is considered in the abovementioned publications. In this case the pressure on the front face of the block is given by

$$P = W_0(1 - 1/n) \tag{4}$$

A generally accepted interpretation of this result is the following. The MDF of the light in free space is equal W_0. The mechanical MDF that enters the block is given by Eq. (4). Then MDF of the light inside the block is equal to W_0/n. As is noted by authors of [3]," only the conservation of momentum and the uniform motion of the center of mass-energy are used, and it is difficult to see how any components of our derivation could seriously be open to question". Indeed, Eq.(4) derived for the pulse where $\tau \ll T$ is not open to question. However, a conclusion that the MDF of the light inside the block is equal to W_0/n is open to question when forces arising in regions of the block where leading and trailing edges of the pulse are propagating are taken into account. These forces are not taken into account because it was supposed that the pulse enters the block instantly. A propagation of one photon is considered usually.

There is another thought experiment [14]. A continuous light wave is reflecting in serial from two parallel reflectors of an optical resonator located in free space. Block in Fig.1 is imbedded in the plane optical resonator. It is shown that a net force applied to the block is equal zero. The pressure on the front face of the block produce by a travelling light wave is given by

$$P_M = W_0(1 - n) . \tag{5}$$

The pressure on the back face is equal $-P_M$. In this case the block is expanded by the pressures only. An increase of the MDF at the entrance of a continuous light wave into the block is explained as follows. There should be a counterpart of the negative pressure P_M applied to the front face of the block. This counterpart is positive and is applied to the object that is responsible for appearance of pressure P_M. Only the continuous light wave can be this object. The positive counterpart increases the MDF of the continuous light wave from W_0 to nW_0.

4. Resolution of a contradiction between Balazs's and Jones's experiments

We have to admit that the MDF of a light pulse decreases in matter by n times but the MDF of a continuous light wave increases in matter by n times. A difference is connected with edges of the light pulse that are absent in a continuous light wave [17]. Let us assume that there are additional pressures in the regions where the leading and trailing edges of the light pulse are propagating. A joint action of the additional pressure produced by the leading edge along with the negative pressure given by Eq. (5) results the positive pressure given by Eq. (4). In this case a magnitude of this additional pressure is given by

$$P_A = W_0(n - 1/n) \qquad (6)$$

Analogously, the additional pressure produced by the trailing edge of the pulse should be equal to $-P_A$. In this case a process of propagation of a light pulse through the block in the Balazs thought experiment at $\tau < T$ looks like as follows. When only the leading edge of the pulse is propagating inside the block, there are two pressures applied to the block. Negative pressure given by Eq. (5) is applied to the front face of the block. Positive pressure given by Eq. (6) is applied to the region where the leading edge is propagating. Time instants when these pressures are terminated are identical and are equal $t = \tau$. As a result, a total pressure on the block is given by $W_0(1-n) + W_0(n-1/n) = W_0(1-1/n)$. This is in accordance with Eq.(4). Thus, the pressure applied to the block obtained from the Balazs thought experiment can be obtained on an assumption that additional pressure in accordance with Eq. (6) takes place in the region where the leading edge of the pulse is propagating. Unlike the interpretation of the Balazs thought experiment that the pressure on the front face of the block is positive and is given by Eq.(4), there is the negative pressure given by Eq.(5). Additional pressure P_A given by Eq. (6) should be taken into account to obtain the pressure in accordance with Eq. (4).

When the trailing edge enters the block, the negative pressure in accordance with Eq. (5) disappears. In the same time the negative additional pressure $-P_M$ in accordance with Eq.(6) in the region where the trailing edge is propagating arises. A sum of pressures produced by the leading and trailing edges of the pulse is equal to zero and the center of mass of the block moves uniformly.

When the leading edge leaves the block, positive pressure P_M in accordance with Eq. (5) arises on the back face of the block. In the same time positive pressure in the region where the leading edge is located disappears but the negative additional pressure produced by the trailing edge of the pulse remains. As a result, a sum of pressures is negative and is equal $-W_0(1-1/n)$. This pressure provides a negative acceleration to the center of mass of the block. The center of mass stops when the trailing edge leaves the back face of the block. This picture is in a full compliance with results of the Balazs thought experiment. As is seen no notion about the Abraham or Minkowski momentums has been used.

Thus, two possible interpretations of the Balazs thought experiment are possible. The first one is generally accepted. The momentum of the light pulse is decreased inside the block by n times. The second one is the following. The momentum of the light pulse is increased by n times but additional pressures in accordance with Eq.(6) arise in the regions where leading and trailing edges of the light pulse are propagating.

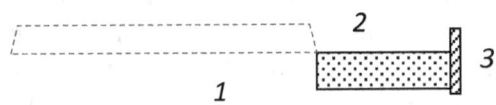

Fig. 2 Propagation of the light pulse 1 (dashed trapeze) in optical medium 2 and subsequent reflection from reflector 3

To choose a correct interpretation, let us consider one more thought experiment where experimental evidence is available. Let a light wave of the energy density W_0 be propagating in free space and enter the block at $t=0$ where it be propagating in an optical medium and be reflecting from an ideal reflector located on the back face of the block (Fig. 2). This block is different from the block in Fig. 1. The slab with gradually decreasing refractive index on the back face of the block is absent.

Let us first consider a propagation of **continuous light wave** in the block. In accordance with the first alternative, the momentum flux density of the continuous light wave inside the block is equal W_0/n, the pressures applied to the front face, to the reflector, and total pressure applied to the block are given by $2W_0(1-1/n)$, $2W_0/n$ and $2W_0$, respectively.

In accordance with the second alternative these pressures are given by $2W_0(1-n)$, $2W_0n$ and $2W_0$. A difference between these alternatives is the following. In accordance with the first alternative all pressures are positive. The pressure applied to the reflector is equal to double MDF that is reflected from the reflector. Then the pressure is equal $2W_0/n$ for the first alternative and $2W_0n$ for the second one. In accordance with the Jones experiment the pressure on the reflector produced by the light propagating in matter is greater by n times than the pressure on the reflector produced by the same light propagating in free space. Thus, the first alternative contradicts to results of the Jones *et al* experiments.

Let us next consider a propagation of **a light pulse** in the block in Fig. 2 to show that the second alternative remains valid, unlike the first one. Let $\tau \gg T$ and the leading and trailing edges be propagating in free space. The leading edge is propagating after reflection in backward direction. The trailing edge is propagating in the forward direction and has not entered the block yet. Since the additional force in free space in accordance with Eq. (6) is equal to zero, this situation is similar to that considered above for a continuous light wave. The total pressure P applied to the system consisting of the block and reflector is equal to a change of the momentum flux density of light in free space per unit time. This change is equal $P=2W_0$. The pressure applied to the front face of the block in accordance with Eq.(5) is equal $P_M=2W_0(1-n)$. Multiplier 2 arises because the pressure P_M is produced by forward and backward waves. From relation $P=P_M+P_R$, where P_R is the pressure applied to the reflector, we have $P_R=2W_0n$. It is not surprising because there is no edges of the pulse within the block and the light pulse can be considered as a continuous light wave in this case.

Let us next assume that $\tau<T$ and $\tau_e \ll \tau$ where τ_e is duration of edges of the pulse. When the leading edge is entering the block, two kinds of pressures arise in accordance with Eq. (5) and Eq. (6). Their sum given by Eq. (4) is consistent with the Balazs thought experiment. When the trailing edge of the pulse enters the block,

the net pressure applied to the block becomes equal to zero because the negative pressure P_M in accordance with Eq. (5) disappears and the positive additional pressure P_A produced by the leading edge in accordance with Eq. (6) is compensated by the negative pressure $-P_A$ produced by the trailing edge.

When the pulse is reflecting from the reflector, the MDF of the pulse changes its direction and pressure $P_R = 2W_0n$ is applied to the reflector in time interval τ. The additional pressures in the regions where the leading and trailing edges of the pulse are propagating are both negative at reflection and their sum is equal to $-2W_0(n-1/n)$. As a result, the total momentum density transferred by the light pulse to the block when the pulse enters the block, propagates inside the block, reflects from the reflector, propagates in backward direction, and leaves the block is equal to $\tau W_0[(1-n)+(n-1/n)+2n-2(n-1/n)+(n-1/n)+(1-n)]=2\tau W_0$. As might be expected, the total momentum density transmitted to the system comprising of the block and reflector is independent on n and is equal to a change of the momentum density of light pulse propagating in free space $2\tau W_0$. The momentum density transmitted to the reflector is equal $2\tau n W_0$. The momentum density transferred to the block is equal $2\tau W_0(1-n)$. A sum of momentum densities transmitted to reflector and block is equal $2\tau W_0$. If a binding between the reflector and block is absent, the block will move to the left, the reflector will move to the right.

Thus, a light pulse can be imagined as a segment of a continuous light wave where additional interaction between light and matter takes place in the regions where edges of the light pulse are propagating. The total MDFs in the light wave and pulse are identical and are equal to W_0n. The positive pressure in the region where the leading edge is propagating produces positive mechanical MDF $W_0(n-1/n)$ in the matter. The negative counterpart of this pressure decreases the electromagnetic MDF of the pulse from nW_0 to $nW_0 - W_0(n-1/n)=W_0/n$. The negative pressure in the region where the trailing edge is propagating produces negative mechanical MDF $-W_0(n-1/n)$ in the matter. The negative and positive MDFs are deleted mutually in the regions located behind the light pulse. Thus, the propagation of the light pulse in matter is accompanied by propagation of two kinds of the momentum. The electromagnetic momentum of density $\tau W_0/n$ and the mechanical momentum density $\tau W_0(n-1/n)$. Their sum is equal $\tau n W_0$ and corresponds to J-momentum. As was shown, the pressure produced by the light pulse on the reflector is equal $2nW_0$. Thus, the mechanical component of the total momentum takes part in producing the pressure on the reflector.

The mechanical component of the total momentum produces the displacement of the block that can be calculated from the following proportion. Pressure in accordance with Eq. (4) causes a displacement of the block Δz given by Eq.(1). Then the pressure given by Eq. (6) causes the following displacement $\Delta z_1=\Delta z(n-1/n)/(1-1/n)=\Delta z(n+1)$. The pressure produces the displacement not only the center of mass of the block but also the whole block because all parts of the block are displaced at identical distance. The negative pressure applied to the front face of the block causes the negative displacement given by $\Delta z_2=\Delta z(1-n)/(1-1/n)=-\Delta zn$. As is seen $\Delta z_1+\Delta z_2=\Delta z$. Although the negative displacement of the center mass of the block is equal Δz_2 due to the pressure given by Eq. (5), the displacement of the regions near the front face of the block is significantly greater. The displacement of the regions near the rear face of the block is equal to zero because the displacement arising near the front face is propagating at sound speed that is smaller by five orders of magnitude than the light speed. The transient processes will be terminated when the light pulse has propagated at great distance above $10^5 D$.

The additional pressures in accordance with Eq. (6) are not taken into account in the generally accepted interpretation of the Balazs thought experiment. Because of this an erroneous conclusion that the momentum of pulse decreases in matter was derived.

As is seen, a transmission of the momentum to the block differs essentially from the simplest view accepted in the generally accepted interpretation of the Balazs thought experiment where it is supposed that the momentum of the pulse simply decreases by n times at entering the block and recovers its value at exiting the block. In reality, it is a complex procedure where the propagation is accompanied by various forces arising in various regions of space in various time instants. These forces produce various mechanical momentums of different signs in different regions of the block. The light pulse transmits to the block mechanical momentums of different signs and leaves the block at light speed. In this time mechanical transient processes initiated by OIF are not terminated although the displacement of the center of mass of the block is equal Δz when the trailing edge of the light pulse has leaved the block. The inner forces that take part in the transient processes change positions of various parts of the block but do not change the displacement of the center of mass.

Thus, a joint consideration of all OIFs enables us to match rival results of the Balazs and Jones experiments. No notion about the Abraham or Minkowski momentum of light is used to interpret a behavior of the block derived from the Balazs thought experiment as well as the pressure produced on a reflector in the Jones *et al* experiments.

5. A physical origin of pressures arising at a propagation of a light in matter

Let us first analyze an origin of pressure on the front face of the block given by Eq. (5). As is shown in Fig.1, there is a slab of thickness d where the refractive index is changed gradually along the z axis from 1 to n. The density force inside the slab is given by

$$f = dP / dz = -W_0 dn / dz = -dn / dz \varepsilon_0 E_0^2 / 2 . \qquad (7)$$

It is desirable to express f in term of the strength of the electrical field in the plane where the density force is determined rather than in term of the strength E_0 in free space. Since there are no reflections inside the slab, the energy density of a continuous light wave in the slab $W = n^2 \varepsilon_0 E^2 / 2$ is greater by n times than energy density $W_0 = \varepsilon_0 E_0^2 / 2$ in free space, where E and E_0 are the electrical fields in the slab and free space, respectively. In this case $E_0^2 = nE^2$. Then Eq.(7) can be presented as follows

$$f = -n\frac{dn}{dz}\varepsilon_0 E^2 / 2 = -\frac{d\varepsilon}{dz}\varepsilon_0 E^2 / 4 \qquad (8)$$

On the other hand, it is known from electrostatics since time of Maxwell [18, 19] that density force in a dielectric located in a constant electrical field E is given by

$$f_{ES} = -grad(\varepsilon)\varepsilon_0 E^2 / 2. \tag{9}$$

As is seen, force f_{ES} is proportional to the square of the electrical field E. In this case force f_{ES} is different from zero in an alternate electrical field also, in particular, in an electrical field of light wave. Averaging f_{ES} over period of oscillation, we obtain Eq.(8) from Eq. (9). Thus, the density force that increases the MDF of the light entering the block through the slab is a counterpart of the force f_{ES} that is known for a long time in electrostatics. Brevik calls this kind of force by the Abraham-Minkowski force [20]. Kemp calls this kind by the Helmholtz force [10]. However this name is more suitable for internal OIF arising due to an inhomogeneity of the electrostriction pressure [21]. As a kind of force f_{ES} was known much earlier, we will call it by the Maxwell-like force because Maxwell was the first who had studied it.

Thus, OIF calculated on the basis of the electrostatic approach in accordance with Eq. (9) (ES approach) is identical to that calculated on the basis of the approach where a change of the momentum flux density is analyzed (CM approach) [14].

Let us next analyze an origin of pressures in the regions where the leading and trailing edges of a light pulse are propagating. The pressure in the region where the leading edge is propagating is given by Eq. (6). The density force produced the pressure on assumption that the energy density W is changed from 0 to W_0 is given by $f_A = (n-1/n)dW/dz$. Since $z=tc/n$, we have $f_A = \dfrac{(n^2 - 1)}{c} \dfrac{dW}{dt}$. Since there are no reflection at the entrance of the light pulse into the block, we have $W=S/c$, where S is the energy flux density $S = [E \times H]$. Thus, we have

$$f_A = \frac{(n^2 - 1)}{c^2} \frac{d[E \times H]}{dt} \tag{10}$$

Density force in Eq.(10) we will call by A-force. It turns out that the A-force is expressed by the same formula as the Abraham force that is considered along with the Abraham momentum in the

Abraham form of the energy-momentum tensor formalism [22, 23]. Usually the term "Abraham force" is used along with term "Abraham momentum". A combination of the Minkowski form of the MDF and the Abraham form of the density force for matching rival experiments causes misunderstanding and objection. What is why we purposely do not use well-known terms of the energy-momentum tensor formalism and use notions from mechanics only. Kinds of force and momentum we denote by terms taken from names of researches who these kinds studied.

Thus, an attempt to coordinate results of unambiguous thought experiments leads to the need to recognize the existence of the optical pressure arising in the regions of an optical medium where the intensity of light is changed in time.

A sum of forces given by Eqs. (9) and (10) gives the following expression for OIF produced by the light, which intensity is changed in time

$$f = -grad(n^2)\varepsilon_0 \frac{E^2}{2} + \frac{(n^2-1)}{c^2} \frac{d[E \times H]}{dt} \qquad (11)$$

Eq. (11) has long been known [22, 23] and used at present time [20].

Thus, our explanation corresponds to the experiments. An inclusion of the A-force into consideration of propagation of light pulses in the Balazs thought experiment enables us to match rival results obtained from the Balazs and Jones experiments. No notions about the Abraham and Minkowski momentums of light are required in this case.

6. Discussion

The A-forces are not taken into account in the generally accepted interpretation of the Balazs thought experiment. Because of this an erroneous conclusion that the momentum of pulse decreases in matter was derived. In reality, the total momentum of the light pulse corresponds to the J-form $W_0 n$. Due to the A-force in the regions where leading and trailing edges are propagating the total MDF is divided into two components: the mechanical component $W_0(n-1/n)$ and the electromagnetic components W_0/n that is equal numerically to B-form. The pressure in accordance with Eq.(5) causes a negative displacement of the center of mass of the block. The pressure in accordance with Eq.(6) causes a positive displacement of the center of mass of the block. A sum of these displacements is positive and corresponds to Eq. (1).

One can see that the J-form of the momentum in matter takes place at a steady state where a change of the light intensity in time is absent. The electromagnetic component is equal to the total MFD nW_0; the mechanical component is equal to zero. This situation takes place in most experiments.

The total MFD of a light pulse in matter is also increases by n times. The electromagnetic component of the light pulse in matter is equal $M_{EM}=W_0/n$. The mechanical component is equal $M_{Mech}=W_0(n-1/n)$. But this is not a single variant.

Let us consider the light pulse which MDF in free space is equal to W_0 at $t<0$ and is equal $W_0+\Delta W$ at $t>0$. Since there is a continuous light wave at $t<0$, we have $M_{total}=W_0n$, $M_{Mech}=0$, $M_{EM}=W_0n$. Then at $t>0$ we have $M_{total}=(W_0+\Delta W)n$, $M_{Mech}=\Delta W(n-1/n)$, and, therefore, $M_{EM}=W_0n+\Delta W/n$. In this case $M_{total}/M_{EM}=(1+\Delta Wn/W_0)/[1+(\Delta W/n)/W_0]$. As is seen, $M_{total}/M_{EM}=1$ at $\Delta W=0$ and $M_{total}/M_{EM}=n^2$ at $\Delta W>>W_0$. Thus, the electromagnetic component of MDF can be an arbitrary part in interval $1...n^2$ of the total MDF. However, this part is not a matter of taste [6] but is determined by a specific situation.

Let us compare distributions of OIF in matter in particular cases on the basis of our notion and notions used in reviews and tutorials. Mansuripur describes an exchange between momentums of light pulse $\mathcal{E}_{pulse}/c$ in free space that enters the block with the refractive index n through anti-reflection $\lambda/4$ plate [8]. Here $\mathcal{E}_{pulse}$ is the energy of the light pulse. He believes that the mechanic and electromagnetic components of the light pulse are equal respectively $(n-1/n)\mathcal{E}_{pulse}/(2c)$ and $\mathcal{E}_{pulse}/(nc)$. The total momentum of the pulse is equal to their sum $(n+1/n)\mathcal{E}_{pulse}/(2c)$. Besides, the mechanical momentum $-(n-1)^2\mathcal{E}_{pulse}/(2nc)$ is transmitted to $\lambda/4$ plate when the pulse enters the block. A sum of this momentum and the total momentum propagating in the block is equal to the momentum of the pulse in free space $\mathcal{E}_{pulse}/c$. As was noted, the law of the conservation of the momentum holds.

However, since MDF of the pulse in accordance with Mansuripur is equal to $W_0(n+1/n)/2$, the pressure on the reflector is not proportional to the refractive index n that contradicts to results of the Jones experiment. Our total MDF in matter is equal to $W_0 n$ and, therefore, the pressure on the reflector is proportional to the refractive index n that agrees with results of the Jones experiment.

Differences between Mansuripur and our analysis are the following. Our total momentum of the pulse, unlike $(n+1/n)\mathscr{E}_{pulse}/(2c)$, is equal to $n\mathscr{E}_{pulse}/c$. Our mechanical momentum, unlike $(n-1/n)\mathscr{E}_{pulse}/(2c)$ is equal $(n-1/n)\mathscr{E}_{pulse}/c$. At last, our mechanical momentum transmitted to $\lambda/4$ plate, unlike $-(n-1)^2\mathscr{E}_{pulse}/(nc)$ is equal $(1-n)\mathscr{E}_{pulse}/c$.

As is seen, a magnitude of the mechanical component of MDF in accordance with Mansuripur $W_0(n+1/n)/2$ is half of our mechanical MDF given by Eq. (6). It is explained by the fact that MDF is produced by the A-force given by Eq. (10) that can be presented as follows

$$f_1 = \dot{P} \times B + P \times \dot{B} \tag{12}$$

where E and B are respectively the electric field strength and the magnetic induction associated with the light radiation, $P=(\varepsilon-1)E$ is the polarization of the optical medium.

OIF produced by the Lorentz force is presented by the second term in the right-hand part of known expression for the Lorentz density force [4, 6, 8, 10] given by

$$f_2 = (P \cdot \nabla)E + \dot{P} \times B \tag{13}$$

On assumption that the leading edge of the light pulse is located in region $z=0...z=l$, we obtain that the pressure produced by the leading edge can be calculated in accordance with Eq. (12) as follows

$$
\begin{aligned}
\boldsymbol{P} &= \frac{\varepsilon\mu-1}{c^2}\int_0^l (H_y\frac{\partial E_x}{\partial t} + E_x\frac{\partial H_y}{\partial t})dz \\
&= \frac{\varepsilon\mu-1}{c^2}\int_0^l (-H_y\frac{\partial H_y}{\varepsilon_0\varepsilon\partial z} - E_x\frac{\partial E_x}{\mu_0\mu dz})dz \\
&= -\frac{\varepsilon\mu-1}{2}\int_0^l (\frac{\mu_0\partial H_y^2}{\varepsilon\partial z} + \frac{\varepsilon_0\partial E_x^2}{\mu dz})dz \\
&= -\frac{\varepsilon\mu-1}{2}\int_0^l (\frac{\mu_0\partial H_y^2}{\varepsilon\partial z} + \frac{\varepsilon_0\partial E_x^2}{\mu dz})dz \\
&= \frac{\varepsilon\mu-1}{2\varepsilon\mu}(\mu_0\mu H_y^2 + \varepsilon_0\varepsilon E_x^2) \\
&= \frac{\varepsilon\mu-1}{\varepsilon\mu}(W_H + W_E) = \frac{\varepsilon\mu-1}{\varepsilon\mu}W_{TOTAL} \\
&= (\sqrt{\varepsilon\mu} - 1/\sqrt{\varepsilon\mu})W_0
\end{aligned}
$$

$$(14)$$

where W_{TOTAL} is the total energy density in matter, W_0 is the total energy density of the same radiation in free space. Since $\varepsilon=n^2$, Eq. (6) is identical to Eq.(14) at $\mu=1$.

As is seen from derivation of Eq.(14), the pressure produced by the Lorentz density force in accordance with second term in the right-hand part of Eq.(13) is smaller by two times because second terms in brackets of Eq.(14) do not take part in calculation of the Lorentz force.

To the best of our knowledge, Jones in 1978 [25] was the first who concluded that the total momentum of light pulse in matter is equal pn, the electromagnetic momentum is equal to p/n and the mechanical momentum is $[1-1/n^2]$ part of the total momentum pn, where p is the momentum of the light pulse in free space. However, he can not explain a physical phenomenon responsible for a rise of the mechanical momentum. He wrote "We are not able to specify the details how this body impulse is created, but merely point out it is demanded by the simple consideration of mechanics".

Mansuripur in 2010 [8] was the second who showed that the mechanical momentums are created by leading and trailing edges of the light pulse. But he used an approach based on the Lorentz density force and his magnitude of the total momentum of the light pulse in matter contradicts to result of Jones *et al* experiment.

We are third who have disclosed a physical origin the Jones "bodily impulse", corrected calculation of Mansuripur and showed that the A-force is responsible for a rise of the mechanical momentum in matter rather than the Lorentz force. The main difficulty was to persuade that an approach based on the Lorentz force is inconsistent in spite of the fact that it is used everywhere in last 40 years.

7. Conclusion

The following conclusion can be derived from matching unambiguous results of thought and real experiments known for a long time. A main reason for conclusion that the momentum of light in matter decreases by n times is an erroneous interpretation of results of the Balazs thought experiment because the A-force arising at leading and trailing edges of a light pulse was not taken into account. An inclusion of these forces enables one to match the results of two rival experiments without introducing a notion about rival forms of the momentum. These forces produce the mechanical momentum distributed between leading and trailing edges of the pulse. The mechanical momentum is propagating together with the light pulse at speed c/n. The B-momentum in a pure form that is derived from the Balazs thought experiment does not exist. The electromagnetic momentum density flux W_0/n is accompanied by the mechanical momentum density flux $W_0(n-1/n)$. Their sum $W_0 n$ corresponds to the J-form. Besides there is the mechanical momentum density flux $W_0(1-n)$ arising at an entrance of the light pulse in matter. This momentum is not propagating together with the light pulse.

The momentum of a continuous light wave corresponds to the J-momentum. The mechanical momentum in this case is absent. It is not surprising. When a continuous light wave enters a medium from free space its momentum increases because there is the Maxwell-like force applied to the medium and directed to free space. In accordance with the third Newton law there is a counterpart of this force applied to the light wave that increases its momentum.

No notion about the B-momentum is required for matching unambiguous results of rival experiments. If the A-force is taken into account, a reason for introducing a notion about the B-momentum disappears completely. In reality, there is one form of the momentum in matter consisting of the electromagnetic and mechanical components. A ration between these components depends on a specific case but a sum of these is greater by n times than that of the same light in free space.

Moreover, we can say that we have derived an existence of a kind of optically induced forces called by the A-force from two unambiguous thought experiments where a light pulse [13] and continuous light wave [14] are analyzed. Only laws of mechanics were used for this purpose. A-force and J-momentum is sufficient for matching rival experiments. We can conclude that the age-old dilemma about a correct magnitude of the momentum of light in matter is solved. A correct magnitude of the total momentum in matter is presented. It is shown that a distribution of the total momentum between its electromagnetic and material components is not a matter a taste as is described in [6] but the distribution can be calculated in each specific case.

References

[1] H. Minkowski, Nachr. Ges. Wiss. Gott, 53, (1908).

[2] M Abraham, Rend. Circ. Mat. Palermo **28**, 1, (1909).

[3] S. M. Barnett and R. Loudon, Phil. Trans. R. Soc. **A 368** 927 (2010).

[4] R. N. C. Pfeifer, T A Nieminen, N.R. Heckenberg and H. Rubinsztein-Dunlop, Rev.Mod.Phys. **79(4)** 1197 (2007).

[5] C. Baxter, R. Loudon, J. of Modern Opt. **57**, 830 (2010).

[6] D.J. Griffiths Am. J. Phys. **80** 7 (2012).

[7] S.M. Barnett, Phys. Rev. Lett. **104**, 070401 (2010).

[8] M. Mansuripur, Opt. Comm. **283** 1997 (2010).

[9] V.P. Torchigin and A.V. Torchigin, Phys. Rev. A **89**, 057801 (2014).

[10] B. A. Kemp, Journal of Applied Physics **109** 111101 (2011).

[11] R. V. Jones and J. C. S. Richard Proc. R. Soc. London Ser. A **221**, 480 (1954).

[12] R. V. Jones and B. Leslie, Proc. R. Soc. London Ser. A **360**, 347 (1978).

[13] N.L. Balazs Phys. Rev. 91, 408 (1953).

[14] V. P. Torchigin and A. V. Torchigin, Annals of Physics **327** 2288 (2012).

[15] R.P. Feynman, R.B. Leighton, M. Sands, *The Feynman lectures on physics,* San Francisco, Addison-Wesley, 2005).

[16] L.D. Landau, E.M. Lifshitz, *Mechanics*. Vol. 1 (Butterworth Heinemann, 1976).

[17] V. P. Torchigin and A. V. Torchigin, Optik **125** 2687 (2014).

[18] J. D. Jackson, *Classical electrodynamics*, (Wiley, N.Y., 1999).

[19] J. A. Stratton, *Electromagnetic theory* (Mc.Graw-Hill, N.Y., 1941).

[20] I. Brevik Phys. Rev. Lett. **103**, 219301 (2009).

[21] V. P. Torchigin and A. V. Torchigin, Opt. Comm. **301-302** 147 (2013).

[22] W. K. H. Panofsky and M. Phillips *Classical electricity and magnetism* (Addison-Wesley, Mass., 1960).

[23] R. N. C. Moller, *The theory of Relavity*, 2nd ed. (Clarendon Press, Oxford, 1972), ch.7.

[24] M. Mansuripur, Optics Expr. **12** 5375 (2004).

[25] R. V. Jones, Proc. R. Soc. London Ser. A **360**, 365 (1978).

Pressure exerted on a semi-infinite lossless dispersionless dielectric by a plane electromagnetic wave

V. P. Torchigin, A.V. Torchigin

Institute of Informatics Problems, Russian Academy of Sciences, Nakhimovsky prospect, 36/1, Moscow, 119278, Russia, tel +7 499 1332532, email: v_torchigin@mail.ru

PACS: 42.65.Jx, 42.65.Tg

Keywords: optical pressure; momentum density; momentum flux; Lorentz force, Abraham force.

Abstract

We show that at present a correct answer to the simplest question about the pressure exerted on semi-infinite transparent lossles dispersionless dielectric at normal incidence of a continuous plane electromagnetic wave is absent at present. It is believed since 1904 that the pressure is negative and is given by -$2W_0(n-1)/(n+1)$ where W_0 is the momentum flux in free space, n is the refractive index of the medium. It is shown in 2002 that the pressure should be positive in accordance with the approach based on the Lorentz density force that is believed at present as a principal theoretical tool that is directly related to the various observable quantities in a transparent way. We use results of two unambiguous but contradictory experiments as a ground of our consideration and conclude that the Lorentz density force approach is incorrect and Abraham density force should be used rather than the Lorentz one. The pressure produced by a continuous wave is negative and coincides with the pressure derived in 1904. The pressure produced by a pulse of wave is the same by value but opposite by sign.

Introduction

It is well known since time of Maxwell that the optical pressure exerted by a light wave propagating in free space on a mirror is equal $2W_0$ where $W_0 = \varepsilon_0 E_0^2/2$, W_0 and E_0 are the momentum flux density and strength of the electrical field of the light wave, respectively. Seemingly, the pressure on a semi-infinite transparent optical medium is also long known. However, a magnitude of the pressure is debated to date. The problem was considered by Poynting [1, 2] many years ago. His calculation predicts an outward force normal to the surface in agreement with previous work by Thomson [3].

Relatively recently in 2002 the optical pressure has been determined on the basis of the approach based on Lorentz force in the simplest configuration at normal incidence of a light wave from free space on a plane boundary of an optical medium. It is shown that the pressure is directed inwards [4], unlike previous opposite opinion derived in 1904 [3]. As is noted [4], it is clearly unsatisfactory that even the direction of the pressure is uncertain for the simplest form of light beam with a plane wave form. Since then all old results were corrected in accordance with the Lorentz force approach. At present, authors of most articles concerned an interaction between light and matter mark as a dignity that they have used the Lorentz force approach.

However, there are doubts that the approach based on the Lorentz force is correct and the corrections are valid. Despite the fact that such statements undermine the foundations of modern ideas, we were able to convince the reviewers of the correctness of our notions and to publish papers where we showed that the approach based on the Lorentz force is incorrect in the simplest situations where a magnitude of the optical pressure is known in advance [5-8].

Certainly, there should be reasonable unambiguous grounds for consideration of this issue. Our grounds are based on two known unambiguous but contradictory experiments.

First, Jones *et al* [9] showed in 1954 that the pressure exerted by the light on a reflector immersed in a liquid is greater by n times, to precision about 1%, as compared with the pressure exerted on the reflector by the same light propagating in a free space where n is the reflective index. The same fact was supported experimentally in 1978 to precision about 0.05% [10]. These results indicate that the momentum of light in matter increases by n times as compared with the momentum of the same light in free space.

Second, these results contradict to results obtained in 1953 by Balazs from the thought experiment based on the generally accepted law of the momentum conservation [11]. It is shown theoretically that an initially motionless transparent block through which a light pulse is propagating without reflection should be replaced in a direction of a propagation of the light pulse if the light pulse enters the block from free space. As a result, the block is moving along the light pulse when the pulse is propagating inside the block. In this case a part of the momentum of the pulse is transferred to the block. Therefore, the momentum of light inside the block is smaller than that of the same pulse propagating in free space. Thus, the momentum of light inside the block corresponds to the Abraham form. As is pointed in review [12], "If argument advanced in favor of the Abraham momentum were to be incorrect, than that would bring into question uniform motion of an isolated body as expressed in the Newton's first law of motion". A description of this experiment is presented in many recent publications [11-14].

Thus, we are based on these two unambiguous facts from mechanics and make use no assumptions about a nature of the optically induced forces and their origin. We show that the facts can be matched if the Abraham density force arising in the regions of the optical medium where leading and trailing edges of the light pulse are propagating is taken into account instead of the Lorentz force. The paper consists of two sections. In the first one we justify grounds derived from unambiguous experiments. Having based on these grounds, we calculate optical pressure produced by a continuous light wave and light pulse on a semi-infinite dielectric in the second section.

2. Resolution of contradiction between the Balazs and Jones experiments as a base for further calculation of the pressure on free-space surface.

Let us first determine notions and laws that are used in the further consideration. If a continuous plane light wave in free space is considered, its momentum is equal to infinity. In this case the momentum density given by $g=[E^{\times}H]/c^2=\varepsilon_0 E^2/c$ is considered where E is the strength of the electrical field. The electrical field of the light wave is changed harmonically as $E_0 \mathrm{Sin}\omega t$. Then the momentum density averaged over period of oscillation is given by $\langle g \rangle = \varepsilon_0 E_0^2/(2c) = W_0/c$ where $W_0 = \varepsilon_0 E^2/2$ is the energy (electrical and magnetic) density of the light wave averaged over period of oscillation. The momentum flux of the continuous plane light wave in free space is well known and is given by $(W_0 c)/c = W_0$ [J/m^3]. A mechanical pressure P applied to a body transmits to the body the mechanical momentum flux equal P [N/m^2=J/m^3].

As follows from the Balazs thought experiment, the momentum of the transparent block where the light pulse is propagating inside the block is given by [12]

$$p_{block} = p_0(1 - 1/n) \tag{1}$$

where p_0 is the momentum of the pulse in free space. As is noted by authors of [12], "only the conservation of momentum and the uniform motion of the center of mass-energy are used, and it is difficult to see how any components of our derivation could seriously be open to question".

Let us consider a unit cross-section of a plane light pulse of duration τ and the energy density W_0 that is propagating in free space along the z-axis and enters the front face of the block at $z=0$ and $t=0$. The momentum of the light pulse in free space is equal to $W_0 c\tau/c = W_0\tau$. The momentum of the block in accordance with Eq. (1) is changed from 0 to $W_0\tau(1-1/n)$ in time interval $0\ldots\tau$. In this case the pressure on the front face of the block when only the leading edge of the pulse is entering the block is given by

$$P = W_0(1 - 1/n) \tag{2}$$

In accordance with the Jones $et\ al$ experiments [9, 10] the momentum flux in matter increases by n times and, therefore the pressure on the front face of the block in given by

$$P_M = W_0(1 - n). \tag{3}$$

There is another thought experiment [5] based on the energetic approach where no assumption about kinds of optically induced force (OIF) is made. A continuous light wave is reflecting in serial from two parallel reflectors of a plane optical resonator located in free space. The transparent block used in the Balazs thought experiment is imbedded in the resonator. Forces applied to the block and the reflector are calculated by means of analysis of a change of total energy stored in the oscillating system at replacement of these objects. Adiabatic invariant $\mathcal{E}/\omega$ is used for the analysis of the change of the energy where $\mathcal{E}$, and ω are the energy stored in the system and its eigen frequency. Well-known methods developed in optics for planes resonators are used for calculation of the change of ω. There is no mention about OIF. It is shown that a net force applied to the block is equal zero and the pressure on the front face of the block produced by a travelling continuous light wave is given by Eq. (3). The pressure on the back face is equal $-P_M$. In this case the block is expanded by the pressures only. Thus, the thought experiment [5] can be considered as the second argument in favor of Eq. (3).

We can conclude from the Balazs and Jones experiments or from Eq.(2) and (3) the following. A joint action of the additional pressure produced by the leading edge when only the leading edge is entering the block along with the negative pressure given by Eq. (3) should give the positive pressure given by Eq. (2). In this case a magnitude of this additional pressure should be equal

$$P_A = W_0(n - 1/n) \tag{4}$$

The density force on the leading edge of the pulse where the energy density W is changed from 0 to W_0 is given by $f_A = dP_A/dz = (n-1/n)dW/dz$ where the z-axis is directed along a direction of propagation of the light pulse and perpendicular to the faces of the block. Since $z=tc/n$, we have $f_A = \dfrac{(n^2-1)}{c}\dfrac{dW}{dt}$. There is no reflection at the entrance of the light pulse into the block. In this case the energy flux density $<S>$ inside the block and free space is identical and is given by $<S>=Wc$, where symbol $< >$ means an average over period of oscillation. Since $\langle S \rangle = \langle [E \times H] \rangle$, we have

$$f_A = \frac{(n^2-1)}{c^2}\frac{d\langle [E \times H] \rangle}{dt}. \tag{5}$$

Eq.(5) coincides completely with expression proposed by Abraham in 1909 for OIF that arises due to a time-rate-change of the intensity of light propagating in an optical medium [15]. Unlike Minkowski who assumed that the momentum flux density in an optical medium increases by n times [16], Abraham assumed that the momentum flux density in an optical medium decreases by n times but at the same time an additional force arises which value coincides with that given by Eq. (5). We have shown on grounds of unambiguous results of two thought experiments [5, 12] that the results can be matched on assumption that the total momentum of light in an optical medium corresponds to the Minkowski form in accordance with Eq. (3). In the same time a time-rate-change of the intensity of light in an optical medium is accompanied by a rise of the Abraham force in accordance with Eq.(5). Thus, an attempt to coordinate results of unambiguous thought and real experiments leads to the need to recognize that the Abraham force is responsible for the pressure produced by a light pulse. An existence of this force is discussed for a long time [17-19]. It is important to underline that no preliminary assumptions about a nature and a physical origin of the optical pressure are used either for Eq.(2) or for Eq.(3) and, therefore, for Eqs. (4) and (5) .

Eq.(3) can be derived alternatively from the following well-known since time of Maxwell [17-19] expression for density force in an inhomogeneous dielectric located in static electrical field of strength E

$$f_{ES} = -grad(\varepsilon)\varepsilon_0 E^2 / 2.$$

(6)

As is seen, f_{ES} is proportional to the square of the electrical field E^2. In this case f_{ES} is different from zero in an alternate electrical field of light wave. In this case Eq. (6) can be used for calculation of optically induced force at a steady-state [5]. Since the law of electrostatics is used in this case, this approach for calculation of optically induced force we will call electrostatic one (ES approach). Let us calculate on the basis of the ES approach the force applied to the front face of the block in the Balazs thought experiment. There is antireflection ●/4 plate on the front face of a transparent block to exclude reflection from the face. Refractive indexes at boundary $z=0$ are equal to $n^{1/2}$ and n, at boundary $z=-$●/4 are equal to 1 and $n^{1/2}$. The energy density of a continuous light wave in the block $W = n^2\varepsilon_0 E^2 / 2$ is greater by n times than energy density $W_0 = \varepsilon_0 E_0^2 / 2$ in free space, where E and E_0 are the electrical fields in the block and free space, respectively. In this case $E^2=E_0^2/n$

In accordance with Eq.(6) pressures averaged over time applied to the boundaries at $z=-\bullet/4$ and $z=0$ are given by

$$P(-\lambda / 4) = -\frac{1}{4}\varepsilon_0 E_0^2 (n-1) \qquad \text{and} \qquad P(0) = -\frac{1}{4}\varepsilon_0 \frac{E_0^2}{n}(n^2 - n),$$

respectively. As is seen, these pressures are identical and their sum is given by Eq. (3). Thus, Eq. (6) can be considered as the third argument in favor of validity of Eq.(3). As a kind of force f_{ES} was considered by Maxwell, we will call it by the Maxwell-like force.

3. Pressures produced by a light on a semi-infinite optical medium

Let the momentum flux of light in free space be W_0, where $W_0=\varepsilon_0 E_0^2/2$ is the energy density of the light in free space. In accordance with the Jones experiment the momentum flux of the light wave penetrating in the medium increases by n times. Let us calculate the pressure produced by a **continuous light wave** on the boundary of the optical medium by means of analysis of a change of the momentum (CM approach)

The energy flux density of the incident wave in free space is equal W_0c. A part of this light wave is reflected. The energy flux density of this part is equal to R^2W_0c. Another part penetrates into the medium. The energy flux density of this part is equal to T^2W_0c where

$$R = \frac{1-n}{1+n}, \quad T = \frac{2\sqrt{n}}{1+n}. \tag{7}$$

As is seen $R^2 + T^2 = 1$. However, the momentum flux in the medium increases by n times. The pressure produced on the medium is equal to a difference between input and output momentum fluxes. In this case we have

$$P = \Delta I = |I_0| + |I_R| - |I_T| = W_0[1 + (\frac{1-n}{1+n})^2 - \frac{4n^2}{(1+n)^2}] =$$

$$-2W_0\frac{n-1}{n+1} = -\varepsilon_0 E_0^2\frac{n-1}{n+1}$$

$$(8)$$

where I_T, I_R and I_0 are momentum fluxes of transmitted, reflected and incident waves, respectively. As is seen, pressure P is negative. Thus, Eq. (8) is consistent with results of Thomson in 1904 [3] but contradicts result of Loudon in 2002 [4].

It is interesting to note that at $n \gg 1$ we have $R=-1$, like $R=-1$ for a mirror. However, the pressure on the mirror is positive and is equal $2W_0$, but pressure on the medium in accordance with Eq. (8) is negative and is equal approximately $-2W_0$ at $n \gg 1$. It is explained by the fact that the momentum flux of the radiation penetrated into the medium increases by n times.

Let us express pressure P in Eq. (8) through amplitude E_B of the electrical field on the boundary. The electrical field is parallel to the boundary and is continuous on the boundary. Since E_B is equal to a sum of electrical fields of the incident and reflected waves, we have

$$E_B = E_0 + E_R = E_0(1 + \frac{1-n}{1+n}) = E_0\frac{2}{1+n}$$

Then Eq.(8) can be presented as follows

$$P = -\varepsilon_0 E_0^2(\frac{n-1}{n+1}) = -\varepsilon_0 E_B^2(\frac{1+n}{2})^2(\frac{n-1}{n+1}) = -\varepsilon_0 E_B^2(n^2 - 1)/4$$

$$(9)$$

The same result follows from Eq.(6) known from time of Maxwell.

$$P = -\varepsilon_0(\varepsilon - 1)E_B^2/4 \qquad (10)$$

Thus, calculation by means of an analysis of a change of the momentum fluxes on assumption that the momentum flux in matter increases by n times as compared with that in free space (CM approach) gives the same result as the generally accepted ES approach. The ES approach is used widely in practical applications, unlike CM approach the use of which is limited by doubts what form of the momentum ought to be used.

Let us now calculate the same pressure produced by a **light pulse** when only the leading edge of the pulse is entering the matter. In this case the additional positive Abraham pressure $(1-R^2)W_0(n-1/n)$ should be added to the pressure given by Eq. (8). Then we have

$$P = -2W_0 \frac{n-1}{n+1} + W_0(1-R^2)(n-1/n) =$$

$$2W_0 \frac{n-1}{n+1} = \frac{n-1}{n+1}\varepsilon_0 E_0^2$$

(11)

Thus, the pressure of a light pulse on a matter when the light pulse is entering the matter from free space is positive, unlike a continuous light wave which produces the same pressure but an opposite sign. Thus, a sign of the pressure produced by light on free-space surface of an optical medium depends on whether a continuous light wave or a light pulse is considered.

Let us calculate for the interest the pressure exerted by a continuous light wave on the same semi-infinite dielectric by means of the Lorentz force approach. The following formula for the Lorentz density force is used

$$\boldsymbol{f}_1 = (\boldsymbol{P} \cdot \nabla)\boldsymbol{E} + \dot{\boldsymbol{P}} \times \boldsymbol{B}$$ (12)

where E and B are respectively the electric field strength and the magnetic induction associated with the light radiation, $P=(\varepsilon-1)E$ is the polarization of the optical medium [14].

In accordance with Eq.(12) the pressure on the boundary is equal zero. Indeed, the first term in Eq.(13) is equal to zero because its z-component is absent. The second term is equal zero in free space where permittivity $P=0$ because $\varepsilon=1$. Phases of P and B of the travelling light wave that penetrates into the medium are identical. In this case phases of dP/dt and B are differed by $\pi/2$ and, therefore, their product averaged over period of oscillations is equal to zero. Thus, in accordance with the Lorentz force approach the pressure exerted by a continuous light wave on a semi-infinite optical medium is equal to zero.

4. Conclusion

From analysis of unambiguous results of the Balazs and Jones experiments a conclusion is derived that there are the Abraham density force arising in the regions of an optical medium where leading and trailing edges of the pulse are propagating. If the Abraham force is taken into account instead the Lorentz force, we obtain that optical pressures produced by a continuous light wave and light pulse on the simplest traditional free-space optical medium are identical by value but opposite by signs. The pressure produced by a continuous light wave is negative and is directed into free space. The pressure produced by a light pulse is positive and is directed from free space until only the leading edge of the pulse is propagating in the medium. When the trailing edge has entered the medium, the pressure becomes equal to zero.

References

[1] J. H. Poynting, Phil. Mag., 9, (1905) 393.

[2] J. H. Poynting, The Pressure of Light (London: Society for Promoting Christian

Knowledge, 1910).

[3] J. J. Thomson, Electricity and Matter (London: Constable, 1904).

[4] R. Loudon, J. of Mod. Opt., 49 (2002) 821.

[5] V. P. Torchigin and A. V. Torchigin, Annals of Physics **327** (2012) 2288.

[6] V. P. Torchigin and A. V. Torchigin, Opt. Comm., **301-302** (2013) 147.

[7] V. P. Torchigin and A. V. Torchigin, Optik **124** (2013) 5492.

[8] V. P. Torchigin and A. V. Torchigin, Phys. Scr. **88** (2013) 035402.

[9] R. V. Jones and J. C. S. Richard, Proc. R. Soc. London Ser. A 221 (1954) 480.

[10] R. V. Jones and B. Leslie, Proc. R. Soc. London Ser. A **360**, (1978) 347.

[11] N.L. Balazs, Phys. Rev. **91** (1953) 408.

[12] S. M. Barnett and R. Loudon, Phil. Trans. R. Soc. **A 368** (2010) 927.

[13] C. Baxter, R. Loudon, J. of Modern Opt. **57**, 42 (2010) 830.

[14] M. Mansuripur, Opt. Comm. **283** (2010) 1997.

[15] M. Abraham, Rend. Circ. Mat. Palermo **28**, 1, (1909).

[16] H. Minkowski, Nachr. Ges. Wiss. Goett, Math. Phys. Kl. 53 (1908).

[17] W. K. H. Panofsky and M. Phillips *Classical electricity and magnetism* (Addison-Wesley, Mass., 1960).

[18] R. N. C. Moller, *The theory of Relativity*, 2nd ed. (Clarendon Press, Oxford, 1972).

[19] I. Brevik, Phys. Rev. Lett. **103** (2009) 219301.

Optical pressure on free-space surfaces

V. P. Torchigin, A.V. Torchigin

Institute of Informatics Problems, Russian Academy of Sciences, Nakhimovsky prospect, 36/1, Moscow, 119278, Russia, tel +7 499 1332532, email: v_torchigin@mail.ru

PACS: 42.65.Jx, 42.65.Tg

Keywords: optical pressure; momentum density; momentum flux density; Lorentz force; Abraham force.

Abstract

We show that generally accepted notions about a magnitude of the optical pressure in the simplest cases when a travelling light is incident normally on a plane boundary between free space and a lossless semi-infinite optical medium should be revised. The pressure is negative for a continuous light wave and is positive for a light pulses unlike zero pressure in accordance with the approach based on the Lorentz density force that is now considered as a principal theoretical tool that is directly related to the various observable quantities in a transparent way. Our consideration is based on grounds derived from unambiguous but contradictory results of two known experiments. No advance assumptions of kinds and a physical origin of optically induced force responsible for the optical pressure are made. We show that the approach based on the Lorentz force is incorrect because its consequence is contrary to the experiments. It is shown that the Abraham density force should be used rather than the Lorentz one. Pressures produced by a continuous light wave and light pulse on various free-space surfaces are presented.

Introduction

It is known since time of Maxwell that the light produces the pressure on a reflecting surface and, therefore, has the momentum. The optical pressure exerted by light in reflection from a mirror in free space was measured long ago [1]. The pressure exerted by the light wave on the mirror is equal $2W_0$ where $W_0 = \varepsilon_0 E_0^2/2$, W_0 and E_0 are the momentum flux and strength of the electrical field of the light wave, respectively. In the same time a magnitude of the pressure on a semi-infinite optical medium is debated to date. The problem was considered by Poynting [2, 3] many years ago. His calculation predicts an outward force normal to the surface in agreement with previous work by Thomson [4]. In the special case of normal incidence, the Poynting result can be presented as follows

$$P = -2W_0 \frac{n-1}{n+1} = -\varepsilon_0 E_0^2 \frac{n-1}{n+1}$$

A situation is changed since 1973 when Gordon proposed an alternative approach based on a belief that the Lorentz density force is responsible for the light pressure [5]. A question about the momentum of light in matter does not arise in this case. Since then, dozens of publications have appeared where optical pressures have been calculated for various configurations of an optical medium. Several reviews have recently appeared [6-10]. It is underlined that the method of calculation based on the Lorentz force avoids any *a priory* expressions for photon momenta [8]. It is believed "that the Lorentz force provides the fundamental description of radiation pressure effects. It has the advantage that no prior assumptions are made about the magnitude of the optical momentum in the medium. Moreover, the calculated force represents the measured quantity, and not some subsidiary quantity that may not itself be directly measurable".

Relatively recently in 2002 the optical pressure has been determined on the basis of the Lorentz force in the simplest configuration where a light wave is incident normally from free space on a plane boundary of a semi-infinite optical medium. It is shown that the pressure is directed inwards [11], unlike previous opposite opinion since 1904 [4]. As is noted [11], it is clearly unsatisfactory that even the direction of the pressure is uncertain for the simplest form of light beam with a plane wave form. Since then all old results were corrected in accordance with the Lorentz force approach. At present, authors of most articles mark as a dignity that they have used the Lorentz force approach.

However, there are doubts that the approach based on the Lorentz force is correct and the corrections are valid. Despite the fact that such statements undermine the foundations of modern ideas, we were able to convince the reviewers of the correctness of our notions and to publish papers where we showed that that the approach based on the Lorentz force is incorrect in the simplest situations where a magnitude of the optical pressure is known in advance [12-15].

Now we present further development of our understanding based on two unambiguous but contradictory results known since 1954 regarding the pressure produced on an optical medium by light propagating from free space into the optical medium. A main part of the paper is devoted to the justification of our ground based on two unambiguous facts known since 1954.

First, Jones *et al* [16] showed in 1954 that the pressure exerted by the light on a reflector immersed in a liquid is greater by n times, to precision about 1%, as compared with the pressure exerted on the reflector by the same light propagating in a free space where n is the reflective index. The same fact was supported experimentally in 1978 to precision about 0.05% [17]. These results indicate that the momentum of light in matter increases by n times as compared with the momentum of the same light in free space.

Second, these results contradict to results obtained in 1953 by Balazs from the thought experiment based on the generally accepted law of the momentum conservation [18]. It is shown theoretically that an initially motionless transparent block through which a light pulse is propagating without reflection should be replaced in a direction of a propagation of the light pulse if the light pulse enters the block from free space. As a result, the block is moving along the light pulse when the pulse is propagating inside the block. In this case a part of the momentum of the pulse is transferred to the block. Therefore, the momentum of light inside the block is smaller than that of the same pulse propagating in free space. Thus, the momentum of light inside the block corresponds to the Abraham form. As is pointed in review [8], "If argument advanced in favor of the Abraham momentum were to be incorrect, than that would bring into question uniform motion of an isolated body as expressed in the Newton's first law of motion". A description of this experiment is presented in many recent publications [7, 8, 16, 19].

Thus, we are based on these two unambiguous facts and make use no assumptions about a nature of the optically induced forces and their origin. We show that the facts can be matched if the Abraham density force arising in the regions of the optical medium where leading and trailing edges of the light pulse are propagating is taken into account instead the Lorentz force. In this case all known old results about pressures produced by light in various situations [6-11, 20-25] obtained on the basis of the Lorentz force approach should be revised.

The paper consists of three sections. In the first one we justify grounds derived from unambiguous experiments. In the second section we calculate optical pressure produced by a continuous light wave and light pulse on the simplest free-space surfaces. In the third part a criticism of the Lorentz force approach is presented.

2. Resolution of contradiction between the Balazs and Jones experiments as a base for further calculation of the pressure on free-space surfaces.

As follows from the Balazs thought experiment, the momentum of the transparent block where the light pulse is propagating inside the block is given by [4]

$$p_{\text{block}} = p_0(1 - 1/n) \tag{1}$$

where p_0 is the momentum of the pulse in free space. As is noted by authors of [8], "only the conservation of momentum and the uniform motion of the center of mass-energy are used, and it is difficult to see how any components of our derivation could seriously be open to question".

If a continuous plane light wave in free space is considered, its momentum is equal to infinity. In this case the momentum density given by $g=[E^{\times}H]/c^2=\varepsilon_0 E^2/c$ is considered where E is the strength of the electrical field. The field of the light wave is changed harmonically as $E_0\sin\omega t$. Then the momentum density averaged over period of oscillation is given by $<g>=\varepsilon_0 E_0^2/(2c)= W_0/c$ where $W_0=\varepsilon_0 E^2/2$ is the energy (electrical and magnetic) density of the light wave averaged over period of oscillation. The momentum flux density (momentum flux) of the continuous plane light wave in free space is given by $(W_0 c)/c = W_0$ [J/m^3]. We will call this momentum flux by the electromagnetic momentum flux because no material objects in free space take part in its production. A mechanical pressure P applied to a body transmits to the body the mechanical momentum flux equal P [N/m^2=J/m^3].

There are optically induced forces (OIF) produced by the light propagating in an optical medium. As a result, the light interacts with matter (an exchange of the momentums between the light and matter takes place). The law of the conservation of the momentums and the third Newton law are valid at this interaction. As a result, each OIF changes the momentum of matter or produces the mechanical momentum flux of matter. In turn, a counterpart of the OIF (COIF) that arises in accordance with the third Newton law changes the electromagnetic momentum flux of light. Each interaction is accompanied a redistribution between mechanical and electromagnetic momentum fluxes. A sum of these momentum fluxes is not changed. Thus, a propagation of light in matter is accompanied by the two components. OIF is responsible for a change of the mechanical component and COIF is responsible for a change of the electromagnetic one. Usually, relations between electromagnetic and mechanical components before their interaction are known. The mechanical component of any light wave in free space is equal zero. In this case, having known a distribution of OIF in space and time, a behavior of the mechanical and electromagnetic components in space and time can be calculated.

Let us consider a unit cross-section of a plane light pulse of duration τ and the energy density W_0 that is propagating in free space along the z-axis and enters the front face of the block at $z=0$ and $t=0$. The momentum of the light pulse in free space is equal to $W_0 c \tau/c = W_0 \tau$. The momentum of the block in accordance with Eq. (1) is changed from 0 to $W_0 \tau(1-1/n)$ in time interval $0...\tau$. In this case the pressure on the front face of the block is given by

$$P = W_0(1 - 1/n) \tag{2}$$

In accordance with the Jones *et al* experiments the momentum flux in matter increases by n times and, therefore, the pressure on the front face of the block in given by

$$P_M = W_0(1-n).$$ (3)

There is another thought experiment [12] based on the energetic approach where no assumption about kinds of OIF is made. A continuous light wave is reflecting in serial from two parallel reflectors of a plane optical resonator located in free space. The transparent block used in the Balazs thought experiment is imbedded in the resonator. Forces applied to the block and the reflector are calculated by means of analysis of a change of total energy stored in the oscillating system at replacement of these objects. Adiabatic invariant $\mathcal{E}/\omega$ is used for the analysis of the change of the energy where $\mathcal{E}$, and ω are the energy stored in the system and its eigen frequency. Well-known methods developed in optics for planes resonators are used for calculation of the change of ω. There is no mention about OIF. It is shown that a net force applied to the block is equal zero and the pressure on the front face of the block produced by a travelling continuous light wave is given by Eq. (3). The pressure on the back face is equal $-P_M$. In this case the block is expanded by the pressures only. Thus, the thought experiment [12] can be considered as the second argument in favor of Eq. (3).

We have to admit that the momentum flux of a light pulse decreases in matter by n times but the momentum flux of a continuous light wave increases in matter by n times. The first attempts to resolve the contradiction was undertaken by Jones [26]. He supposed that mechanical processes should accompany a propagation of a light pulse in matter and these processes should be connected with the mechanical momentum that is replaced together with the light pulse. Mansuripur assumed that the mechanical processes are produced by the Lorentz density force arising in regions where leading and travelling edges are propagating [19]. However, this assumption is incorrect because, as is shown below, the Abraham force should be used rather than the Lorentz force.

We can conclude from the Balazs and Jones experiments or from Eq.(2) and (3) the following. A joint action of the additional pressure produced by the leading edge when only the leading edge is entering the block along with the negative pressure given by Eq.

(3) should give the positive pressure given by Eq. (2). In this case a magnitude of this additional pressure produced by the leading edge of the pulse should be equal

$$P_A = W_0(n - 1/n) \tag{4}$$

The density force on the leading edge of the pulse where the energy density W is changed from 0 to W_0 is given by $f_A = dP_A/dz = (n-1/n)dW/dz$ where the z-axis is directed along a direction of propagation of the light pulse and is perpendicular to the faces of the block. Since $z = tc/n$, we have $f_A = \dfrac{(n^2 - 1)}{c}\dfrac{dW}{dt}$. There is no reflection at the entrance of the light pulse into the block. In this case the energy flux density $<S>$ inside the block and free space is identical and is given by $<S> = Wc$, where symbol $<\ >$ means an average over period of oscillation. Since $\langle S \rangle = \langle [E \times H] \rangle$, we have

$$f_A = \frac{(n^2 - 1)}{c^2}\frac{d\langle [E \times H] \rangle}{dt}. \tag{5}$$

Eq.(5) determines the Abraham force [27 -31] rather than the Lorentz force. Thus, an attempt to coordinate results of unambiguous thought and real experiments leads to the need to recognize that the Abraham force is responsible for the pressure produced by a light pulse. An existence of this force is discussed for a long time. As is shown, the existence can be derived theoretically from two thought experiments [18, 12]. It is important to underline that no preliminary assumptions about a nature and a physical origin of the optical pressure are used either for Eq.(2) or for Eq.(3)and, therefore, for Eqs. (4) and (5) .

This distinguishes our approach from other recently published approaches where various assumptions about a nature and reasons of a rise of OIF are made [19-25]. For example, it is believed in last 40 years that the Lorentz force is responsible for a rise of OIF. It turns out that a sum of the Lorentz force and another additional kind force gives the Abraham force in a field of electromagnetic waves propagating in an optical medium [14]. The Abraham force is equal to zero at a steady-state. Thus, ought to distinguish optical pressures produced by a continuous light wave and light pulses.

Eq.(3) can be derived alternatively from the following well-known since time of Maxwell [27-33] expression for density force in an inhomogeneous dielectric located in static electrical field of strength E

$$f_{ES} = -grad(\varepsilon)\varepsilon_0 E^2 / 2. \qquad (6)$$

As is seen, f_{ES} is proportional to the square of the electrical field E^2. In this case f_{ES} is different from zero in an alternate electrical field of light wave. In this case Eq. (6) can be used for calculation of optically induced force at a steady-state [12]. Since the law of electrostatics is used in this case, this approach for calculation of optically induced force we will call electrostatic one (ES approach). Let us calculate on the basis of the ES approach the force applied to the front face of the block in the Balazs thought experiment. There is antireflection ●/4 plate on the front face of a transparent block to exclude reflection from the face. Refractive indexes at boundary $z=0$ are equal to $n^{1/2}$ and n, at boundary $z=-$●/4 are equal to 1 and $n^{1/2}$. The energy density of a continuous light wave in the block $W = n^2 \varepsilon_0 E^2 / 2$ is greater by n times than energy density $W_0 = \varepsilon_0 E_0^2 / 2$ in free space, where E and E_0 are the electrical fields in the block and free space, respectively. In this case $E^2 = E_0^2 / n$

In accordance with Eq.(6) pressures averaged over time applied to the boundaries at $z=-$●/4 and $z=0$ are given by

$$P(-\lambda / 4) = -\frac{1}{4}\varepsilon_0 E_0^2 (n-1) \qquad \text{and} \qquad P(0) = -\frac{1}{4}\varepsilon_0 \frac{E_0^2}{n}(n^2 - n),$$

respectively. As is seen, these pressures are identical and their sum is given by Eq. (3). Thus, Eq. (6) can be considered as the third argument in favor of validity of Eq.(3). As a kind of force f_{ES} was considered by Maxwell, we will call it by the Maxwell-like force.

3. Pressures produced by a light on a plane free-space surfaces

Following [11, 21], an ideal submerged reflector, semi-infinite dielectric, slabs of different widths and a slab with anti-reflection $\lambda/4$ coating layers are considered.

Ideal reflector

In accordance with the Jones experiment that is used as a basis of our consideration the pressure exerted by a continuous light wave on a reflector is proportional to the refractive index of an optical medium. In this case the pressure is equal $2nW_0$. If a light pulse is considered, the pressure on the reflector is also equal $2nW_0$. In the same time there are additional pressures in accordance with Eq.(4) in the regions of the optical medium where leading and trailing edges are propagating.

For example, if the leading edge is propagating in the optical medium after reflection from the reflector and the trailing edge is propagating in the optical medium before reflection, they produce on the medium pressure $-2W_0$ $(n-1/n)$. The total pressure produced by the light pulse on the system consisting of the reflector and medium is $2nW_0-2W_0(n-1/n)=2W_0/n$.

This is another example. If both leading and trailing edges of a light pulse are propagating in an optical medium in the same directions, the total pressure produced by the edges is equal to zero. The leading edge produces the positive mechanical momentum in regions where it is propagating. The trailing edge produces the negative mechanical momentum in regions where it is propagating. As a result there is the mechanical momentum only in the regions between the edges that is equal $W_0(n-1/n)$. Thus, the total momentum of the light pulse W_0n consists of the mechanical momentum $W_0(n-1/n)$ and the electromagnetic momentum W_0/n.

Semi-infinite dielectric

Let us first consider the simplest case when a continuous light wave enters normally from free space a homogeneous optical medium with the reflective index n through the plane boundary between free space and the optical medium. Let the momentum flux in free space be W_0, where $W_0=\varepsilon_0 E_0^2/2$ is the energy density of the light wave in free space. In accordance with the presented conception, momentum flux of the light wave penetrating in the medium increases by n times. Let us calculate the pressure produced by the light wave on the boundary of the optical medium by means of analysis of a change of the momentum fluxes (CM approach).

The energy flux density of the incident wave in free space is equal $W_0 c$. A part of this light wave is reflected. The energy flux density of this part is equal to $R^2 W_0 c$. Another part penetrates into the medium. The energy flux density of this part is equal to $T^2 W_0 c$ where

$$R = \frac{1-n}{1+n}, \quad T = \frac{2\sqrt{n}}{1+n}. \tag{7}$$

As is seen $R^2 + T^2 = 1$. However, the momentum flux in the medium increases by n times. The pressure produced on the medium is equal to a difference between input and output momentum fluxes. In this case we have

$$P = \Delta I = |I_0| + |I_R| - |I_T| = W_0[1 + (\frac{1-n}{1+n})^2 - \frac{4n^2}{(1+n)^2}] =$$

$$- 2W_0 \frac{n-1}{n+1} = -\varepsilon_0 E_0^2 \frac{n-1}{n+1}$$

$$\tag{8}$$

where I_T, I_R and I_0 are momentum fluxes of transmitted, reflected and incident waves, respectively. As is seen, pressure P is negative. Thus, Eq. (8) is consistent with results of Thomson in 1904 [4] but contradicts result of Loudon in 2002 [11].

Let us express pressure P in Eq. (8) through amplitude E_B of the electrical field on the boundary. The electrical field is parallel to the boundary and is continuous on the boundary. Since E_B is equal to a sum of electrical fields of the incident and reflected waves, we have

$$E_B = E_0 + E_R = E_0(1 + \frac{1-n}{1+n}) = E_0 \frac{2}{1+n}$$

Then Eq.(8) can be presented as follows

$$P = -\varepsilon_0 E_0^2 (\frac{n-1}{n+1}) = -\varepsilon_0 E_B^2 (\frac{1+n}{2})^2 (\frac{n-1}{n+1}) = -\varepsilon_0 E_B^2 (n^2 - 1)/4$$

$$\tag{9}$$

The same result follows from Eq.(6) known from time of Maxwell.

$$P = -\varepsilon_0(\varepsilon - 1)E_B^2 / 4 \tag{10}$$

Thus, calculation by means of an analysis of a change of the momentum fluxes on assumption that the momentum flux in matter increases by n times as compared with that in free space (CM approach) gives the same result as the generally accepted ES approach. The ES approach is used widely in practical applications, unlike CM approach the use of which is limited by doubts what form of the momentum ought to be used.

Let us now calculate the same pressure produced by a **light pulse** when only the leading edge of the pulse is entering the matter. In this case the additional positive Abraham pressure $(1-R^2)W_0(n-1/n)$ should be added to the pressure given by Eq. (8). Then we have

$$P = -2W_0 \frac{n-1}{n+1} + W_0(1-R^2)(n-1/n) = 2W_0 \frac{n-1}{n+1} = \frac{n-1}{n+1}\varepsilon_0 E_0^2 \tag{11}$$

Thus, the pressure of a light pulse on a matter when the light pulse is entering the matter from free space is positive, unlike a continuous light wave which produces the same pressure but an opposite sign. Thus, a sign of the pressure produced by light on free-space surface of an optical medium depends on whether a continuous light wave or a light pulse is considered.

Eq.(11) agrees with Eq. (53) in [11] where the single-photon momentum transferred into medium is considered. However there is no contradiction between Eq. (53) in [11] derived by Loudon in 2002 and Eq. (1) in [11] derived by Poynting in 1904 because Eq. (53) applies to a light pulse and Eq. (1) applies to a continuous light wave.

Let us next calculate the pressure on the boundary produced by a continuous wave propagating from left to right from an optical media into free space. The pressure is equal the following difference of momentum fluxes

$$P = W_0 n + W_0 nR^2 - W_0(1-R^2) = 2nW_0(n-1)/(n+1).$$ As is

seen, the pressure is positive and is directed towards propagation of light into free space. This also agrees with Eq. (6)

Let us last calculate the same pressure produced by a light pulse. In this case additional negative Abraham pressure takes place $-W_0(n-1/n)$ produced by the trailing edge when the leading edge is propagating in free space. As a result, the total pressure is given by

$$P = 2nW_0(n-1)/(n+1) - W_0(n-1/n) = W_0\{[(n-1)/(n+1)](2n-1-1/n)\} \quad . (12)$$

It turns out that the pressure on an optical medium by a light pulse propagating from the medium into free space normally to the boundary is negative at $n<n_0=1+2^{1/2}$, is equal to zero at $n=n_0$, and is positive at $n>n_0$. The pressure produced on the boundary is positive. The pressure produced on the medium in the region where the trailing edge is propagating is negative. A sign of their sum depends of the refractive index n.

Slab of $\lambda/2$ width

Let us calculate pressures applied to the left-hand and right-hand faces of the slab shown in Fig. 1 as well as a total pressure applied to the slab. The total pressure can be calculated as a difference between input and output momentum fluxes. These flaxes are located in free space and, therefore, a knowledge of the momentum fluxes in an optical medium is not required. Momentum fluxes in the slab take part in production of pressures on the faces of the slab. A sum of these pressures is equal to the total pressure which calculation is not in doubt. Thus, we can check the fact that the pressure on a plane can be determined from a change of the momentum fluxes transmitted through the plane. Let us assume that the incident and reflected wave is given by

$a_0 \exp[i(\omega t - 2\pi z/\lambda_0)]$, $b_0 \exp[i(\omega t + 2\pi z/\lambda_0)]$ in free space and $a_1 \exp[i(\omega t - 2\pi nz/\lambda_0)]$, $b_1 \exp[i(\omega t + 2\pi nz/\lambda_0)]$ inside the slab where λ_0 is the wavelength in free space. It is supposed that dimensionless amplitudes $a_0=1$ corresponds to the energy flux W_0c.

The following relations between waves in Fig. 1 take place

$$a_2 = a_1 T \exp(-i2\pi nL/\lambda_0), \quad b_2 = 0, \quad a_1 = a_0 T - Rb_1,$$
$$b_1 = -Ra_1 \exp(-i4\pi nL/\lambda_0), \quad b_0 = b_1 T + Ra_0$$

$$(13)$$

where L is width of the slab. From these relations we have

$$a_1 = a_0 T / [1 - R^2 \exp(-i4\pi nL / \lambda_0)],$$

$$b_1 = a_0 TR \exp(-i4\pi nL / \lambda_0) / [1 - R^2 \exp(-i4\pi nL / \lambda_0)],$$

$$a_2 = a_0 T^2 \exp(-i2\pi nL / \lambda_0) / [1 - R^2 \exp(-i4\pi nL / \lambda_0)],$$

$$b_0 = -a_0 \{ T^2 R \exp(-i2\pi nL / \lambda_0) / [1 - R^2 \exp(-i4\pi nL / \lambda_0)] - R \}$$

$$(14)$$

For slab of width $L = \lambda_0/2n$ we have $\exp(-i4\pi nL / \lambda_0) = 1$ and

$$a_2 = -a_0, \ a_1 = a_0 / T, \ b_1 = -R/T, \ b_0 = 0, \ b_2 = 0.$$

$$(15)$$

As is seen, there is no reflection from the slab because $b_0=0$. The dimensionless energy flux inside the slab is equal $\left|a_1^2\right| - \left|b_1^2\right| = \left|a_0^2\right|(1/T^2 - R^2/T^2) = \left|a_0^2\right|$. Thus, the energy flux inside the slab is equal to the incident energy flux in free space.

Taking into account that $b_0=0$ and $b_2=0$, the total pressure on the slab is equal

$$P_{total} = W_0(|a_0|^2 - |a_2|^2) = 0. \tag{16}$$

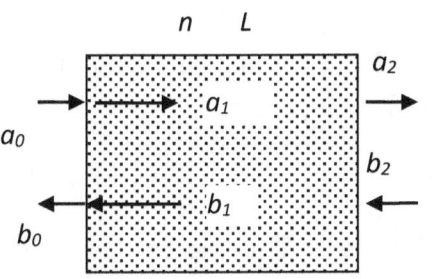

Fig.1. Designation of incident, transmitted and reflected waves in slab

The pressure on the left-hand face of the slab is equal to a difference between a sum of input momentum fluxes $W_0\left(\left|a_0^2\right| - n\left|b_1^2\right|\right)$ and a sum of output momentum fluxes $W_0\left(n\left|a_1^2\right| - \left|b_0^2\right|\right)$. Here, it is supposed that the momentum fluxes in an optical medium increase by n times as compared with momentum fluxes in free space. As a result, we have

$$P_{left} = W_0[(|a_0^2| - n|b_1^2|) - (n|a_1^2| - |b_0^2|)]$$
$$= W_0[(|a_0^2| + |b_0^2|) - (n|b_1^2| + n|a_1^2|)]$$
(17)

Analogously, we have for the pressure on the right-hand of the face is given by

$$P_{right} = W_0[(n|a_1^2| - |b_2^2|) - (|a_2^2| - n|b_1^2|)] =$$
$$W_0[-(|a_2^2| + |b_2^2|) + n(|a_1^2| + |b_1^2|)]$$
(18)

Taking into account Eq. (15), we have
$$P_{total} = P_{left} + P_{right} = W_0(|a_0^2| - |a_2^2| - |b_0^2| + |b_2^2|) = 0$$
$$P_{left} = W_0(1 - nR^2 / T^2 - n / T^2) = W_0(1 - n^2) / 2$$
$$P_{rigth} = W_0(-1 + n / T^2 + nR^2 / T^2) = W_0(n^2 - 1) / 2$$

As is seen from Eqs. (17), (18), P_{total}=0 independently on multiplier n.

Slab of $\lambda/4$ width

In this case we have $L = \lambda_0/4n$. Then from Eq.(14) we have
$$a_1 = a_0 T / (1 + R^2) = \sqrt{n}(1 + n) / (1 + n^2),$$
$$a_2 = a_0 T^2 \exp(-j\pi / 2) / (1 + R^2) = 2n \exp(-j\pi / 2) / (1 + n^2)$$
$$b_0 = 2R / (1 + R^2) = (1 - n^2) / (1 + n^2)$$
(19)

From Eq. (19) we have $|a_2^2| + |b_0^2| = |a_0^2|$. This means that a sum of the energy densities of the transmitted and reflected waves is equal to that of the incident wave. The pressures on the left-hand, right-hand faces and the total pressure on the slab in accordance with Eqs. (17), (18), (19) are, respectively, equal

$$P_{left} = W_0[1 + |b_0^2| - n(|a_1^2| + |b_1^2|)] = -2W_0(n^2 - 1) / (1 + n^2)^2$$
(20)

$$P_{right} = W_0[n(|a_1^2| + |b_1^2|) - |a_2^2|] = 2W_0 n^2 (n^2 - 1)/(1+n^2)^2$$

(21)

$$P_{total} = W_0[|a_0^2| + |b_0^2| - |a_2^2|)] = 2W_0|b_0^2| = 2(1-n^2)^2/(1+n^2)^2$$

(22)

As is seen, $P_{total} = P_{left} + P_{right}$ and in accordance with Eq. (20), (21) P_{left} is negative and P_{right} is positive. This agrees with Eq. (6). Pressures P_{left} and P_{right} were calculated purposely to demonstrate a calculation of pressures by means of analysis of a change of the momentum fluxes taking into account an increase of the momentum flux by n times in an optical medium.

Slab of arbitrary width with two antireflection coating

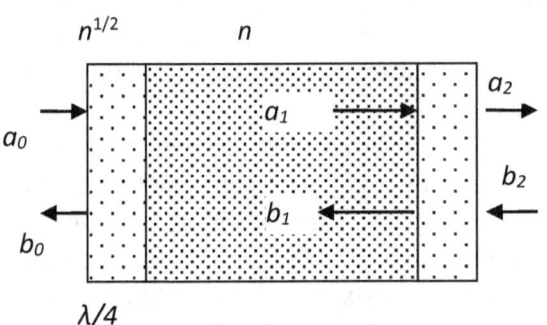

Fig.2. Designation of incident, transmitted, and reflected waves in slab with two antireflection coatings

This slab is used in the Balazs thought experiment. The pressures produced by a continuous light wave and the leading edge of a light pulse on the front face of the slab are given by Eq. (3) and Eq.(2), respectively, that agrees with the contradictory results of the Balazs and Jones experiments. As was shown, an explanation is based on the fact that a continuous light wave is considered in the Jones experiment and a light pulse is considered in the Balazs thought experiment. In the last case the Abraham force arising in the regions where edges of the pulse are propagating should be taken into account.

4. Criticism of the approach based on the Lorentz density

force

Let us compare our results with results obtained on the basis of the Lorentz force approach. The following formula for the Lorentz density force is used

$$f_1 = (P \cdot \nabla)E + \dot{P} \times B \qquad (23)$$

where E and B are respectively the electric field strength and the magnetic induction associated with the light radiation, $P=(\varepsilon-1)E$ is the polarization of the optical medium [19].

In accordance with Eq.(23) the pressure on the boundary is equal zero. Indeed, the first term in Eq.(23) is equal to zero because its z-component is absent. The second term is equal zero in free space where permittivity $P=0$ because $\varepsilon=1$. Phases of P and B of the travelling light wave that penetrates into the medium are identical. In this case phases of dP/dt and B are differed by $\pi/2$ and, therefore, their product averaged over period of oscillations is equal to zero. Thus, in accordance with the Lorentz force approach the pressure produced by a continuous light wave on free space surface of an optical medium is equal to zero and is independent of the refractive index n.

Having known the pressure on the boundary, we can determine the momentum flux in the medium and, therefore, factor α that characterizes an increase of the momentum flux in the matter as compared with that in free space. If the energy flux of the incident wave is equal $S=W_0 c$, the energy fluxes of the reflected and transmitted waves are equal $R^2 S$ and $T^2 S$, respectively, where indexes R and T are given by Eq. (7).

Taking into account that the pressure on a plane is equal to a difference between input and output momentum fluxes, we obtain that the pressure on the boundary between free space and the optical medium is given by

$$P=S/c+R^2S/c-\alpha T^2S/c=0 \qquad (24)$$

and, therefore, $\alpha=(1+R^2)/T^2$. Taking into account Eq. (7), we have

$$\alpha = (n+1/n)/2 \qquad (25)$$

Thus, the momentum flux in an optical medium is the arithmetical mean between Abraham and Minkowski forms of the momentum flux in matter. This result agrees with that of Mansuripur [19, 20], Loudon [11], Loudon and Baxter [11] but contradicts to ours where $\alpha=n$.

Let us next show that the pressure produced on a reflector by a continuous light wave which the momentum flux in matter increase by factor given by Eq.(23) contradicts to result of the Jones experiment. We have shown that results of the Jones experiment are explained naturally because the momentum flux of a continuous light wave in matter increases by n times. It is impossible to explain results of the Jones experiment if the momentum flux increases by $(n+1/n)/2$ times. Mansuripur and Zakharian in article entitled "Whence the Minkowski momentum ? attempted to save the approach. In Appendix of the paper they describe "a method of calculating the mechanical momentum transferred to the host medium as a result of the overlap between incident and reflected pulses in the dielectric region immediately preceding the reflector". They hope that "a net force experienced by the mirror will be greater than that expected from a simple argument based solely on the conservation of the pulse's total (i.e. electromagnetic and mechanical) momentum".

First of all, ought to note, that if the pressure experienced by the mirror is equal to $2nW_0$ in accordance with the Jones experiment, the momentum flux incident into the mirror should be equal nW_0 because the pressure is produced due to a change of the momentum flux.

Let us analyze the regions where the Lorentz force can increase the momentum flux from $W_0(n+1/n)/2$ to nW_0. Let us calculate the total pressure produced by the Lorentz force in the simplest particular case (Fig. 3) where the slab of width $3\lambda/(2n)$ is confined by $\lambda/4$ antireflection plate of width $\lambda/(4n^{1/2})$ that is located in region $-\lambda/(4n^{1/2})\ldots0$. The slab is located in region $z=0\ldots 3\lambda/(2n)$ and is confined by a mirror. The field of a plane standing wave in the slab is given by

$$E_x = 2(E_0/\sqrt{n})Sin(nkz)Cos(\omega t), E_y = 0, E_z = 0$$

$$H_x = 0, H_y = 2(E_0\sqrt{n})Cos(nkz)Sin(\omega t), H_z = 0$$

$$(26)$$

where E_0 is the amplitudes of the strength of electrical field of a travelling wave propagating in free space, $k=\omega/c=2\pi/\lambda$, ω is the angular frequency of the light wave, $Z_0=(\mu_0/\varepsilon_0)^{1/2}$. The electrical and magnetic fields in their maximums are increased by two times as compared with the amplitude of the corresponding travelling wave.

As is seen, the first term in Eq. (23) is equal to zero because in accordance with Eq. (26) the y- and z- components of the electrical field E are absent and $dE_x/dx=0$. Then from Eqs. (23, 26) we have

$$\dot{P}_x = -2\omega\varepsilon_0(\varepsilon-1)\eta^{-1/2}E_0 Sin(knz)Sin\omega t \quad \text{and}$$

$$B_y = 2E_0\eta^{1/2}(\mu_0\varepsilon_0)^{1/2}Cos(knz)Sin\omega t.$$

In accordance with Eq.(23) the z-component of the Lorentz force is given by

$$f_z = \dot{P}_x B_y = -4\varepsilon_0(\varepsilon-1)(\omega/c)E_0^2 Sin(knz)Cos(knz)Sin^2(\omega t).$$

Taking into account that $\omega/c=k=2\pi/\lambda$ and time-averaging f_z over period of oscillation, we obtain the following expression for the density force in a plane resonator

$$f_z = -(2\pi/\lambda)\varepsilon_0(\varepsilon-1)E_0^2 Sin(4\pi nz/\lambda).$$

(27)

The pressure produced by the Lorentz force on the slab is given by

$$P = \int_0^{3\lambda/(2n)} -(2\pi/\lambda)\varepsilon_0(\varepsilon-1)E_0^2 Sin(4\pi nz/\lambda)dz =$$

$$\int_0^{6\pi} -(1/2n)\varepsilon_0(\varepsilon-1)E_0^2 Sin(\zeta)d\zeta = 0$$

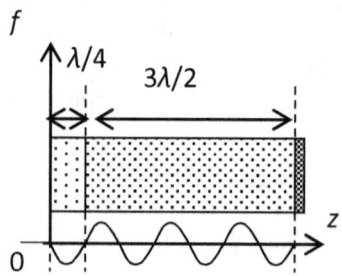

Fig. 3. Distribution of the Lorentz density force f_z along block

The same is seen in Fig.3 where the negative Lorentz density force is compensated by positive one at each region of $\lambda/(2n)$ length. Only pressure produced by the negative Lorentz force applied to $\lambda/4$ plate is not compensated. The amplitude of the electrical field on the left face of $\lambda/4$ plate is equal to $2E_0$ where E_0 is the amplitude of the electrical

144

field in free space on the left of the plate. Taking into account that the refractive index in $\lambda/4$ plate is equal to $n^{1/2}$ and designating $\zeta=4\pi n^{1/2}z/\lambda$, we obtain that the pressure produced on the $\lambda/4$ plate by the Lorentz force in accordance with Eq. (27) is given by

$$P = \int_0^\pi -(1/2\sqrt{n})\varepsilon_0(n-1)E_0^2 Sin(\zeta)d\zeta = -2W_0(\sqrt{n}-1/\sqrt{n})$$

(28)

This pressure is different from pressure given by Eq. (3) that provides an increase of the momentum flux from W_0 in free space to W_0n in matter as it is required by the Jones experiment.

Moreover, the pressure in accordance with Eq. (28) is different from pressure $-W_0(n-1)^2/n$ presented in [21] where the pressure produced by a continuous travelling light wave on $\lambda/4$ plate is calculated. It is explained by the fact that the pressure in [21] is calculated for $\lambda/4$ plate through which a travelling wave is propagating. Thereafter it is supposed that the pressure is increased by two times if the additional wave reflected from the mirror is propagating backwards through the $\lambda/4$ plate. This is true from a general point of view but it is incorrect if the Lorentz force approach is used.

Here are other simplest examples. It is difficult to come up with simpler. We need to explain them in detail because of our numerous unsuccessful attempts to persuade opponents for a long time that the approach based on the Lorentz force is incorrect. Let a plane travelling electromagnetic wave be propagating in a homogeneous optical medium along the z-axis. Let us calculate the pressure applied to fragment of optical medium between planes $z=z_1$ and $z=z_2$. The pressure is equal to zero because the momentum flux densities of incident and transmitted waves are identical. Now, let the same wave be propagating in opposite direction. The pressure exerted by this wave on the fragment is also equal to zero. Let both waves be propagating simultaneously. The waves are produced a standing wave. As was shown above in accordance with Eqs. (20), (21), the pressure produced by waves in any plane is equal to a difference of the momentums fluxes between a sum of incident waves and a sum of reflected and transmitted waves. In the case the difference is equal to zero and, therefore, the total pressure produced by these waves on the fragment is also equal to zero.

Let us calculate the same pressure on the basis of the Lorentz force approach. The Lorentz density force of a travelling wave averaged over period of oscillation $<f_1>$ in accordance with Eq. (23) is equal zero because phases of P and B in the travelling wave are identical and, therefore, phases of dP/dt and B are shifted by $\pi/2$. In this case a product of dP/dt and B averaged over period of oscillation is equal to zero.

Let us next consider pressures in the optical medium located in a field of standing wave. Phases of P and B are shifted by $\pi/2$ in the standing wave. In this case product of dP/dt and B averaged over period of oscillation is different from zero and, therefore, density force $<f_1>$ is different from zero. The pressure produced by the Lorentz density force on the fragment is equal to integral of $<f_1>$ from z_1 to z_2. The integral is different from zero for arbitrary z_1 and z_2 and, therefore, momentum flux densities of the transmitted waves should be changed. Thus, the approach based on the Lorentz force is self-contradictory. On one hand, it requires that the momentum of a travelling light wave propagating in a homogeneous optical medium should not be changed. On other hand, it requires that the momentum of the same travelling light wave propagating in a homogeneous optical medium where other light wave is propagating should be changed.

The same situation connected with an absence of an additively of momentum fluxes of light wave in the Lorentz force approach takes place in the considered above example where a light wave is propagating through $\lambda/4$ plate. Author of [21] considers a travelling wave and obtains that the pressure on the plate is given by $-W_0(n-1)^2/2n$. We consider the standing wave and obtain that the pressure on the plate is given by $-2W_0(\sqrt{n}-1/\sqrt{n})$ rather than $-2W_0(n-1)^2/2n$. Thus, approach based on the Lorenz force gives incorrect results in the simplest cases and a long-term activity based on the approach is senseless.

5. Conclusion

From analysis of unambiguous results of the Balazs and Jones experiments a conclusion is derived that there are the Abraham density force arising in the regions of an optical medium where leading and trailing edges of the pulse are propagating. If the Abraham force is taken into account instead the Lorentz force used in last 40 years, optical pressures produced by a continuous light wave and light pulse on the simplest traditional free-space surfaces should be corrected. Sometimes even a sign of the pressure should be changed.

A dispute about a sign of the optical pressure on a free-space surface of a dielectric by a light incident from free space normally on the surface is resolved as follows. The pressure produced by a continuous light wave is negative and is directed into free space. The pressure produced by a light pulse is positive and is directed from free space until only the leading edge of the pulse is propagating in the medium.

A sign of the light pressure produced by the light pulse propagating from an optical medium into free space depends on the refractive index of the medium.

In our opinion, an activity on the use of the approach based on the Lorentz density force is erroneous.

References

[1] P. Lebedev, Annal. Phys, Leipzig 6, (1901) 433.

[2] J. H. Poynting, Phil. Mag., 9, (1905) 393.

[3] J. H. Poynting, The Pressure of Light (London: Society for Promoting Christian
Knowledge, 1910).

[4] J. J. Thomson, Electricity and Matter (London: Constable, 1904).

[5] J. P. Gordon Phys. Rev. A 8, 14, (1973).

[6] R N C Pfeifer, T A Heckenberg and H. Rubinsztein-Dunlop, Rev.Mod.Phys. **79** (2007) 1197.

[7] C. Baxter, R. Loudon, J. of Modern Opt. **57**, 42 (2010) 830.

[8] S. M. Barnett and R. Loudon, Phil. Trans. R. Soc. **A 368** (2010) 927.

[9] B. A. Kemp, Journal of Applied Physics, **109** (2011) 111101.

[10] D.J. Griffiths, Am. J. Phys. **80** (2012) 7.

[11] R. Loudon, J. of Mod. Opt., 49 (2002) 821.

[12] V. P. Torchigin and A. V. Torchigin, Annals of Physics **327** (2012) 2288.

[13] V. P. Torchigin and A. V. Torchigin, Opt. Comm., **301-302** (2013) 147.

[14] V. P. Torchigin and A. V. Torchigin, Optik **124** (2013) 5492.

[15] V. P. Torchigin and A. V. Torchigin, Phys. Scr. **88** (2013) 035402.

[16] R. V. Jones and J. C. S. Richard, Proc. R. Soc. London Ser. A 221 (1954) 480.

[17] R. V. Jones and B. Leslie, Proc. R. Soc. London Ser. A **360**, (1978) 347.

[18] N.L. Balazs, Phys. Rev. **91** (1953) 408.

[19] M. Mansuripur, Opt. Comm. **283** (2010) 1997.

[20] M. Mansuripur, Opt. Comm. **283** (2010) 3557.

[21] M. Mansuripur, Optics Expr. **12** (2004) 5375.

[22] M. Mansuripur and A. R. Zakharian, Phys. Rev. A 80 (2009) 023823.

[23] A.R. Zakharian, V. Vansuripur, J.V. Moloney, Optics Expr. **133** (2005) 2321.

[24] R. Loudon, S.M. Barnett, C. Baxter, Phys. Rev. A 71 (2005) 063802.

[25] R. Loudon, S.M. Barnett, Opt. Expr. 14 (2006) 11855.

[26] R. V. Jones, Proc. R. Soc. Lond. A 360 (1978) 365.

[27] W. K. H. Panofsky and M. Phillips *Classical electricity and magnetism* (Addison-Wesley, Mass., 1960).

[28] R. N. C. Moller, *The theory of Relavity*, 2nd ed. (Clarendon Press, Oxford, 1972).

[29] I. Brevik, Phys. Rev. Lett. **103** (2009) 219301.

[30] Brevik and S. Ellingsen, Phys. Rev. A 81 (2010) 011806.

[31] I. Brevik and S. Ellingsen, Phys. Rev. A **86** (2012) 025801.

[32] J. D. Jackson *Classical electrodynamics*, (Wiley, N.Y., 1999).

[33] J. A. Stratton *Electromagnetic theory* (Mc.Graw-Hill, N.Y., 1941).